INTRODUCTORY COLLEGE MATHEMATICS

ROBERT D. HACKWORTH, Ed.D.
Department of Mathematics
St. Petersburg Junior College at Clearwater
Clearwater, Florida

and

JOSEPH HOWLAND, M.A.T.
Department of Mathematics
St. Petersburg Junior College at Clearwater
Clearwater, Florida

SAUNDERS SERIES IN MODULAR MATHEMATICS

Metric Measure

W. B. Saunders Company: West Washington Square
Philadelphia, PA 19105

12 Dyott Street
London, WC1A 1DB

833 Oxford Street
Toronto, Ontario M8Z 5T9, Canada

INTRODUCTORY COLLEGE MATHEMATICS ISBN 0-7216-4422-8
Metric Measure

 Made in the United States of America. Press of W. B. Saunders Company. Library of Congress catalog card number 75-23629.

Last digit is the print number: 9 8 7 6 5 4 3 2 1

PREFACE

Metric Measure

This book is one of the sixteen content modules in the Saunders Series in Modular Mathematics. The modules can be divided into three levels, the first of which requires only a working knowledge of arithmetic. The second level needs some elementary skills of algebra and the third level, knowledge comparable to the first two levels. *Metric Measure* is in level 2. The groupings according to difficulty are shown below.

Level 1	Level 2	Level 3
Tables and Graphs	*Numeration*	*Real Number System*
Consumer Mathematics	*Metric Measure*	*History of Real Numbers*
Algebra 1	*Probability*	*Indirect Measurement*
Sets and Logic	*Statistics*	*Algebra 2*
Geometry	*Geometric Measures*	*Computers*
		Linear Programming

The modules have been class tested in a variety of situations: large and small discussion groups, lecture classes, and in individualized study programs. The emphasis of all modules is upon ideas and concepts.

The conversion to the metric system by the United States makes *Metric Measure* an important module for all science and non-science majors. The module is a necessity for students enrolled in science or technology courses in which the metric system is used.

Metric Measure begins by presenting some of the history of measurement while showing some of the difficulties of developing adequate standards of measure and measurement systems. The organization of the metric system is followed by methods of converting measurements from one system or unit to another. The module deals with linear, area, volume, mass, liquid, and temperature measures.

In preparing each module, we have been greatly aided by the valuable suggestions of the following excellent reviewers: William Andrews, Triton College, Ken Goldstein, Miami-Dade Community College, Don Hostetler, Mesa Community College, Karl Klee, Queensborough Community College, Pamela Matthews, Chabot College, Robert Nowlan, Southern Connecticut State College, Ken Seydel, Skyline College, Ara Sullenberger, Tarrant County Junior College, and Ruth Wing, Palm Beach Junior College. We thank them, and the staff at W. B. Saunders Company for their support.

Robert D. Hackworth
Joseph W. Howland

NOTE TO THE STUDENT

Objectives:

Upon completing this unit, the reader is expected to be able to demonstrate the following skills and concepts:

1. An understanding of ancient measurement systems and the difficulties involved in using non-standardized units.

2. An ability to change measures from the inch-pound system to the metric system.

3. An ability to change measures from the metric system to the inch-pound system.

4. An ability to change measures in the metric system from one unit to another.

Three types of problem sets, with answers, are included in this module. Progress Tests appear at the end of each section. These Progress Tests are always short with only four to six problems. The questions asked in Progress Tests always come directly from the material of the section immediately preceding the test.

Exercise Sets appear less frequently in the module. More problems appear in an Exercise Set than in a Progress Test. These problems arise from all sections of the module preceding the Exercise Set. Part I problems in the Exercise Sets are specifically chosen to match the objectives of the module. Part II problems are Challenge Problems.

A Self-Test is found at the end of the module. Self-Tests contain problems representative of the entire module.

In learning the material, the student is encouraged to try each problem set as it is encountered, check all answers, and re-study those sections where difficulties are discovered. This procedure is guaranteed to be both efficient and effective.

CONTENTS

METRIC MEASURE

INTRODUCTION

A recent headline stated, "POLICE SEIZE KILO OF MARIJUANA." The feeling for the amount of marijuana that was seized based on the previous headline may have varied tremendously from person to person. Some people may have thought that a whole roomful of marijuana was found. Others may have had the impression that just a small container was seized. Possibly most readers had no idea how much a kilo weighed. Actually a kilogram weighs approximately 2.2 pounds. For many Americans, the experience of interpreting measurements given in metric units may occur sooner than people realize. The United States may soon adopt a new system of weights and measures, called the metric system, and drop the old system of measuring lengths in inches, feet, yards, and miles, and weights by ounces, pounds, and tons. Hopefully, the change will be gradual -- allowing time for people to learn to "think metric." There are good reasons for believing that the metric system will replace the inch-pound system in the United States. In 1975, the United States was the only industrialized nation in the world not using the metric system. Great Britain, the originator of the English inch-pound system of measure, has adopted metric units as its official system of weights and measures. Canada and Mexico, our neighbors to the north and south, have officially adopted the metric system. The countries of the European Common Market have gone so far as making an agreement to accept only those products that are in metric units and sizes after 1977. Foreign corporations that are building factories in the United States must have employees capable of working with metric units. American corporations currently using inch-pound measurements will have to build at least some of their products to metric units so they can meet world competition. As a country interested in exporting and importing vast quantities of material, the United States will need an increasing number of people with an ability to build and repair products produced to metric specifications. Consequently, the pressures of world trade will force the United States and its individual

---

A recent bill in the Florida legislature proposed that the liter be used as the largest legal size allowable for the sale of alcoholic beverages. A liter bottle holds slightly more whiskey than a quart bottle.

businesses to deal with goods measured and designed in metric units. The scientific community in the United States is already using the metric system. Widespread, everyday use of the metric system cannot be far in the future.

The economic well-being of the United States is an important factor in considering an adoption of the metric system, but another factor may ultimately be far more significant. The metric system is a far simpler system for almost every type of problem involving computation or arithmetic with measurements. There are two reasons for the simplified computations in the metric system. First, metric measurements are given in decimals instead of fractions or mixed numbers. Most people find calculations are easier with decimals than with fractions or mixed numbers. Second, changing units in the metric system is always a case of multiplying or dividing by 10, 100, 1000, etc. Such multiplication and division in a decimal system is relatively simple because it can be accomplished simply by moving the decimal point.

Progress Test 1

1. Which of the following seems to describe the height of a six foot tall man?

   a. 543 cm b. 1.8 m c. 96 mm

2. The girl with the hour-glass figure in a metric country would have measures :

   a. 91-55-91 cm b. 53-34-53 cm

   c. 15-9-15 cm d. 36-26-36 cm

3. Most people have trouble adjusting to change. Did you have trouble with problems 1 and 2 in confronting a new system of measures?

ANCIENT MEASUREMENT SYSTEMS

The history of measurement probably begins about 10,000 BC. There is evidence that the people of that time were using linear measurements. Linear measures are those used for lengths. Inches, feet, and miles are examples of linear measures in the inch-pound system. Centimeters, meters, and kilometers are examples of linear measures in the metric system. In the primitive cultures of the Neolithic Age (10,000 BC to 8,000 BC) linear measures were probably used to compare the lengths of spears, the sizes of animals killed, or the lengths of sticks in a lean-to shelter.

Measures of volume were developed when people learned to farm and cultivate food grains. Tribal chieftains probably set up measuring units of cups, pots, jars, or baskets to measure volumes. There is archeological evidence of devices for measuring volume dating back to 2050 BC. A mural found in an Egyptian tomb shows a copper cylinder for measuring oil and another cylinder for measuring wheat.

Ancient linear measuring units apparently were based on comparisons with parts of the body. One of the most interesting, complete reports written on ancient linear measures was prepared by John Quincy Adams in 1821 when he was United States Secretary of State. Adams listed and described the following linear measures:

cubit -- the distance from the top of the middle finger to the tip of the elbow.

ell -- twice the cubit or the distance from the top of the middle finger to the middle of the breast when the arm is extended horizontally to the side.

fathom-- twice the ell or the distance between the tips of the middle fingers when the arms are extended horizontally.

span -- one-half a cubit.

palm -- one-third a cubit.

finger-- one-fourth a cubit.

foot -- one-sixth of a fathom.

Three interesting facts are illustrated by the list of ancient measuring units. First, the description of the ell makes the tacit assumption that the distance from the elbow to the middle of the breast is equal to one cubit. Second, each of these units is based upon the size of the measurer's body. Third, the subdivision of units was in halves, thirds, fourths, and sixths.

The use of weights and measures preceded recorded history. The origin of weights and measures was attributed to the gods; the Egyptians to their god Theuth and the Greeks to their god Mercury.

Measuring lengths through the use of body parts is a natural method that almost every person has attempted. The fisherman who holds his hands apart to show the "big one that got away" has a long tradition of body-part-measurers behind him. There are certainly advantages to carrying one's measuring stick as part of one's body, but there are also distinct disadvantages. The fellow with long arms has a different set of measuring units than a shorter person. Sometimes that may be an advantage, sometimes a disadvantage, but always it means there is poor comparability.

The subdivision of ancient units into halves, thirds, fourths, and sixths has been carried over to our present inch-pound system. Today, most foot rulers are marked off into twelfths (inches), half-inches, quarter-inches, and eighth-inches. Pounds are divided into sixteenths (ounces). Pints are half-quarts and each pint is divided into sixteenths (ounces again). Apparently these divisions are the result of the subdivisions used in ancient measuring units, but they depart drastically from a base ten number system which is built upon multiples or divisions of 10.

Although the cubit may have been the first unit of linear measuring and some people consider it sacred because Noah supposedly used it in building his ark, the foot became the basic linear unit in the cultures surrounding the Mediterranean from 2000 BC until the advent of the metric system in the 1800's AD. Nevertheless, the foot was a non-standardized unit. Each small geographic area had its own "foot" in terms of length. The table below shows some of these different "foot" measurements.

Culture	Approximate Measurement in Inches
Sumeria	10.50
Greek	12.14
Phoenicia	10.98
Carthage	11.08
Sardinia	11.17
Rome	11.30

As the table clearly shows, the foot varied from culture to culture. The variation was probably due to the size of particular individuals who established the measuring unit. The number of different units was limited only by the number of chieftains for the separate villages, fiefdoms, or nations. Those differences are understandable for a time in history when distances greatly inhibited travel and trade. More surprising perhaps is the fact that even today the United States Coast and Geodetic Survey uses a foot that varies from the standard by 0.000 06 foot.

Measurements for weights were established by 8300 BC. An Egyptian balance from that time has been found which evidently was used for weighing precious metals. Because weight measures originally were intended for grains such as wheat and barley, the "grain" became the unit of weight for precious metals. Today the "grain" is still used by pharmacists, drug manufacturers, scientists, and the government. The amount of gold equal to the United States dollar is measured in grains. Of course, the grain as it is used today is quite accurate and standardized, but the ancient measurement of "grain" varied with the rainfall, grain type, richness of the soil, and the moisture content on a particular day. The shrewd trader may have realized that on hot, humid days the grain was wet and heavy whereas in times of drought or low humidity the grain was light and dry.

Progress Test 2

1. What fraction of a cubit is a foot?

2. Hector and Homer, from two different villages, weighed a gold ring, each man using his own grain for a unit of weight. Hector said the ring weighed 18 grains, but Homer said the ring weighed 24 grains. Which man's grain had the lightest weight per kernel?

3. Suppose that Alitta, a village chieftain, set up the following set of measures for measuring wine.

   cup - the basic unit, jar - ten times as large as a cup

   firkin - ten times as large as a jar

   baryl - ten times as large as a firkin

   a. 3 jars would equal _____ cups.

   b. 3 firkins would be equivalent to _____ cups

c. 3 baryls would be equivalent to _____ cups.

d. True or false. 1.5 firkins equal .15 baryl.

EFFORTS TO STANDARDIZE UNITS

Most of the history of man's measurements is marked by the use of natural non-standardized units such as the length of body parts or the weight of grains of barley. Natural units that were dependent upon a particular time, place, or individual would vary in size. There could be no uniformity of measure with these natural units. The lack of standardized units common to all people who used them was recognized by the Egyptians 4000 years ago. The Egyptians attempted to overcome the lack of standardized units and ease the problems of computation about 2000 BC.

The Egyptians established a basic unit of linear measure. From this basic unit other measuring units were developed. The Egyptians constructed relationships into their measuring system. Other measuring systems of that time, and the present inch-pound system, first named the units and then tried to find relationships between them.

Unfortunately, the Egyptian system was ignored by other cultures and declined in use with the rise of the Roman Empire. The Romans used a natural base measurement system based on twelve subdivisions to a unit. For example, there were 12 unciae (ounces) to a libra (pound). As the Romans conquered nations they made no attempt to force their entire system on the other cultures, but the Romans did insist that the basic units be named by the Roman names and subdivided into twelve units. Therefore, in the Roman Empire every basic unit of weight was called a pound and divided into twelve ounces. Although the pound varied radically in weight from England to Judea, it always contained 12 ounces.

At the end of the 9th Century, Charlemagne, famous for the Magna Carta, attempted to standardize systems of weight and measure throughout his Empire. Two stories are told of the source of his linear foot which was 12.789 3 inches long. One said it was actually the length of Charlemagne's foot. The other story said that it was exactly one-half of the Arabic cubit (established by the Caliph from the "Arabian Nights").

Although the Charlemagne foot existed as the French standard until displaced by the metric system, his edicts on measurements had limited effect. Local war lords ignored them, preferring to use their own personal standards instead. Peasants living far from the seats of power continued to use the units they had inherited from their ancestors. From Charlemagne's time the number of measuring units seemed to constantly expand. An indicator of the number of units used in England alone is a dictionary of the units used from the 12th Century to the 19th Century. The dictionary contains 184 pages of unit names ranging from alna to butt to dicker to pokke to werkop. As an example of the changing standard for just one unit, a barrel contained 32 gallons in the 13th century and was equal to 4 firkins of 8 gallons each. In 1688 it contained 34 gallons, but by 1803 it was changed to 36 gallons. The ale barrel had a capacity of 32 gallons, but a barrel of herrings only contained 30 gallons. Barrels of salmon were the largest at 84 gallons. A barrel of wine held 31-1/2 gallons after 1750. In Ireland, the barrel held 40 gallons of grain. Since the size of the gallon varied just as much as the barrel, it is indeed difficult to know how much a barrel really held.

The Asian countries also followed the natural tendency to use familiar objects as measurement standards.

One ancient Asian country, Thailand, had an interesting way of organizing linear units. In Thailand, small linear units were expressed in terms of hairs, louse eggs, and grains of rice. In early times, one atom would be one-sixty-fourth of a hairbreadth, eight atoms made up one molecule, and eight molecules would be a hairbreadth. Eight hairbreadths were equal to one louse egg, eight louse eggs became one louse, eight lice would be one grain of rice, two grains of rice (why two instead of eight?) made one krabiad, and four krabiads were equal to one fingerbreadth. The Thai people may have used these relatively small linear measurements in the construction of their exquisitely engraved gold and silver jewelry.

One of the familiar nursery rhymes may have resulted from a King's tax on a 17th century unit of volume. Charles I placed a tax on the jack or jackpot, a pot used to hold honey, wine or milk. As a result, the size of the jack was reduced. Half a jack made a gill. There is reason to suspect that the rhyme about Jack and Jill was a protest against the jack tax. Gill or Jill were synonyms for sweetheart. When Jack fell down, Jill came tumbling after.

Progress Test 3

1. Romans had to be familiar with unciae and libras. If 14-1/2 libras are converted into unciae, how many unciae were obtained?

2. 32 unciae = ? libras.

3. According to the Thai measurement system, a line as long as a grain of rice would be _____ hairbreadths long.

4. If a design on a Thai ring was 1 krabiad wide, how many atoms wide would it be?

THE BIRTH OF THE METRIC SYSTEM

As varied and confusing as the chaotic multitude of measurement units used by ancient cultures may seem, the difficulties did not bother most people because of the lack of mobility at that time. Most people never traveled as much as 50 miles from their birthplace. A brewer in Birmingham did not care that a gallon in Hamburg was a different size because he never dealt with the brewers in Hamburg anyway. Traders both tolerated and profited from the different measuring systems when exchanging one unit of merchandise for another. The only people who were starting to see the advantage of a well organized system of weights and measures based on some natural phenomenon were the scientists. This group, exchanging ideas, problems, and solutions on an international scale, was greatly hampered by the lack of common measurement standards. The scientists also wanted a measurement system that facilitated computations.

As chaotic as measurement systems were in England, Europe, and Asia, they were no better in newly established America. Like other countries, the United States, in its early years, used standards for measurement that were natural. Consequently, they varied in size from town to town. In George Washington's time, the length of three barleycorns was equivalent to one inch. Since the length of three barleycorns will vary from farm to farm, one inch in South Carolina might have varied significantly from the length of an inch in Boston.

Thomas Jefferson, in 1784, proposed two workable plans for systems of weights and measures, but the plans did not achieve Congressional support.

One of Jefferson's proposals was to "define and render uniform and stable the existing system...to reduce the dry and liquid measures to corresponding capacities by establishing a single gallon of 270 cubic inches and a bushel of eight gallons or 2 160 cubic inches."

Jefferson's second proposal was interesting because it had some of the features of the metric system. He proposed "to reduce every branch to the same decimal ratio already established for coin and thus bring the calculations of the principal affairs of life within the arithmetic of every man who can multiply and divide plain numbers." Jefferson defined a natural linear unit as "a cylindrical rod of iron of such length as in latitude 45^0 in the level of the ocean and in a cellar or other place, the temperature of which does not vary through the year, shall perform its vibrations in small and equal arcs in one second of mean time." The effects of gravity, air pressure, and temperature were considered by Jefferson's definition. Therefore, it could be duplicated wherever the conditions of latitude, altitude, and temperature could be met.

The other units in Jefferson's system were derived from the basic linear unit. A foot was related to the distance that an iron rod swings in one second. An inch would be one-tenth of a foot. A line would be one-tenth of an inch. A point would be one-tenth of a line. Longer units were: a decade which was ten feet; a rood which was ten decades; a furloung which was ten roods; and a mile which was ten furloungs. Jefferson defined an ounce as "the weight of a cube of rainwater of one-tenth of a foot; or... the thousandth part of the weight of a cubic foot of rain water, weighed at standard temperature."

Sharing the fate of so many worthy ideas, Jefferson's proposals died in the Congress. Only his proposal to decimalize the currency was adopted.

The idea of basing linear units on objects such as a person's foot was abandoned by the French scientists in the 1790's. At that time Talleyrand, a powerful leader in the French National Assembly, proposed that the Academy of Sciences be instructed to study a universal system of weights and measures for France.

The scientists in the Academy insisted that a nonvariable base be used as a standard for linear measure. After considering the distance a pendulum swings in a second and a section of the length of the equator, the Academy decided on a section of the arc of the meridian through Dunkirk and Barcelona on the Mediterrenian Sea. A meridian is a circle around the earth perpendicular to the equator. After the distance from Dunkirk to Barcelona was measured with a great deal of difficulty over a seven year period, that distance was used to fix the length of the meter at one ten-millionth of the distance from the North pole to the equator. The meter was one of the units defined in the new "metric system." Three platinum and several iron bars were constructed by the French to serve as physical representations of the meter. If another country wanted to use the meter as its official standard, it had to have a bar constructed according to the length of the bars in Paris. This awkward and time consuming procedure was necessary to the standardization of the meter.

All the other linear units determined by the Academy were multiples or divisions of the meter by powers of 10. A kilometer is 1 000 times a meter and a centimeter is 1/100 of a meter. Therefore this new system was a decimal system: moving the decimal in a measure created larger or smaller units.

The table below shows the prefixes established for the linear multiples and divisions of the meter.

Myriameter	10 000 meters
Kilometer	1 000 meters
Hectometer	100 meters
Dekameter	10 meters
Meter	1 meter
Decimeter	0.1 meter
Centimeter	0.01 meter
Millimeter	0.001 meter

Each unit in the table above is 10 times longer than the unit in the next line below. A decimeter is 10 times as long as a centimeter. 0.1 = 0.01 x 10. A dekameter is 10 times longer than a meter.

Conversions are not so easy in the English system of weights and measures. The figure on the opposite page compares the conversion difficulties of the metric and inch-pound systems.

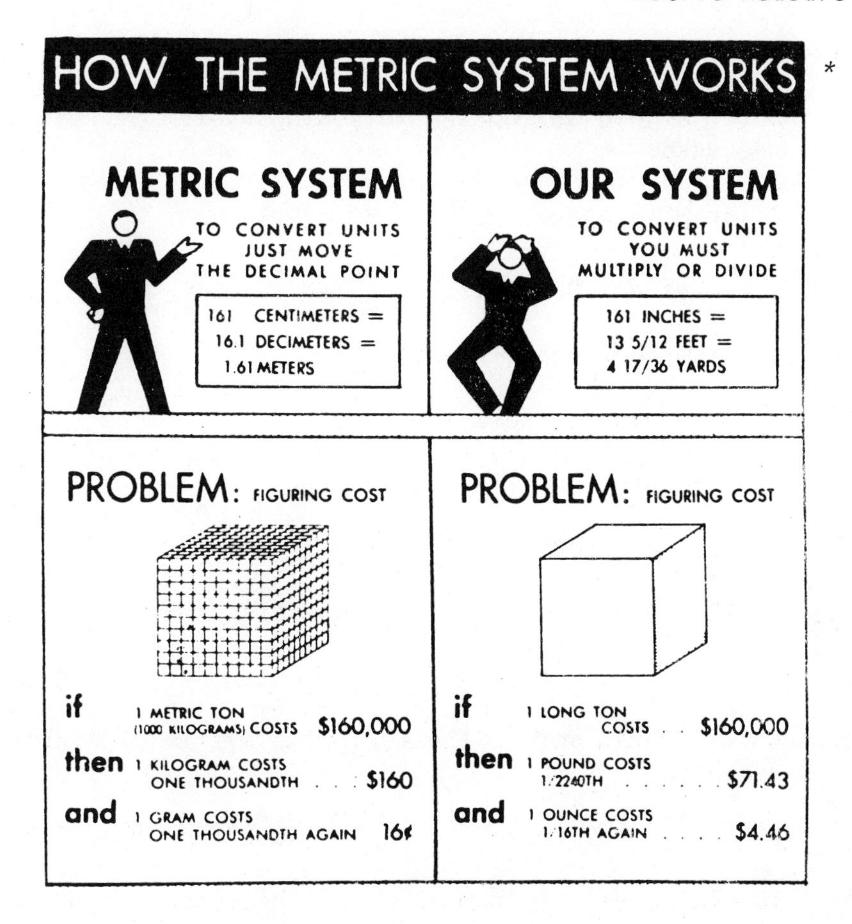

*

Progress Test 4

1. The length of a king's girdle was once used as the basis for the length of a yard. What is wrong with using such a length as a basis for a unit measurement?

2. Place the following units in order from the shortest to the longest length. Centimeter, kilometer, meter, hectometer, millimeter.

3. A kilometer is ________meters.

4. A kilometer is ________hectometers.

5. A millimeter is ________of a centimeter.

---

*From *Twentieth Yearbook of the National Council of Teachers of Mathematics*, published by Bureau of Publications, Teachers College, Columbia University. Copyright 1948.

Exercise Set 1

I. 1. Chief Highiq set up the following system of volume measures.

1 jar = 10 pots   1 cup = $\frac{1}{10}$ pot

1 basket = 1 000 pots

a. How many cups in a pot?

b. How many baskets in a pot?

c. How many cups in a basket?

d. 25 jars contain __________ cups.

e. 3 500 pots = __________ baskets.

f. 0.457 basket = __________ pots.

2. Chief Highiq set up the following system of linear measures.

10 inches = 1 foot   1 decifoot = 10 feet

10 mininches = 1 inch   1 hectofoot = 100 feet

1 kilofoot = 1 000 feet

a. 25 inches = __________ feet

b. The chief's sword was 0.6 decifoot long. How many feet long was it?

c. The chief's wife always had to walk 4.5 decifeet behind him. How many feet behind did she walk?

d. The tribe's war canoe was 0.76 hectofoot long. How many feet long was it?

e. The distance between the chief's eyes was 25.4 mininches. How many inches apart were his eyes?

f. The distance to the next village was 34.3 kilofeet. How many hectofeet was it to the next village?

3. a. A fathom is approximately 6 feet long. How many feet long is an ell?

   b. 1 fathom equals __________ cubits.

   c. If Noah built an ark 100 cubits long, about how many feet long was it?

   d. The length of two fingers is equivalent to how many cubits?

   e. How many cubits high is a horse that measures 15 palms high?

   f. 29 spans = __________ cubits.

   g. 100 fingers = __________ palms.

   h. $\frac{1}{4}$ ell = __________ palms.

4. Chief Lowiq set up the following system of linear measures for his village.

   1 foot = 8 inches   1 mile = 4 800 feet

   1 yard = 5 feet

   Use the chief's system to answer the questions.

   a. 48 inches = __________ feet

   b. 3 feet = __________ inches

   c. The chief's foot was 12 inches long.

      12 inches = __________ feet.

   d. The distance from Chief Lowiq's village to the village of the chief mentioned in problems 1 and 2 was 5 280 feet. 5 280 feet = __________ miles.

   e. Chief Highiq of problems 1 and 2 had a hut 8 feet high using his measuring system. Chief Lowiq's hut was $9\frac{3}{4}$ feet high using his units. Their inch units were the same length. Who had the tallest hut?

5. Give a reason for the fact that the Sumerian foot had the shortest length for its time.

6. Use the Thai system of linear measures to answer the questions.

   a. If a bold broach is 1 fingerbreadth wide, how many atoms wide is it?

   b. A design that is 3 hairbreadths wide is ________ atoms wide.

   c. A king wants a diamond that has a diameter two lice wide. How many hairbreadths wide will it be?

7. Using the length of a portion of a meridian is better than using the length of a person's foot as a standard of measure. Find a disadvantage of using the length of a meridian as a standard of measure.

8. a. 20 000 meters = __________ myriameters.

   b. 2 000 millimeters = __________ meters.

   c. 300 centimeters = __________ meters.

   d. 50 hectometers = __________ meters.

   e. 500 kilometers = __________ meters.

   f. 3 meters = __________ centimeters.

   g. 3 000 meters = __________ kilometers.

   h. 0.5 meter = __________ millimeters.

   i. 2 meters = __________ dekameters.

   j. 5 000 meters = __________ myriameters.

9. In the example titled How the Metric System Works, the answers to the metric problem all contained the three digits 1, 6, and 0. The answers to the problem using our units contained many different digits. Why?

II. Challenge Problems

1. Using the system of measures from problems 1 and 2 of part I;

   a. 0.25 mininch = __________ inches.

   b. 0.085 6 kilofoot = __________ decifeet.

   c. 3 495 hectofeet = __________ mininches.

   d. $\frac{1}{5}$ of a cup = __________ basket.

2. Chief Lowiq and Chief Highiq used the same unit for their inch. $3\frac{1}{2}$ miles to Chief Lowiq is __________ kilofeet to Chief Highiq.

3. It is approximately 4 800 kilometers across the United States.

   a. How many myriameters is it across the United States?

   b. 4 800 kilometers = __________ centimeters.

4. When the metric system is adopted, centimeters will be used as a unit of measure instead of inches.

   1 foot $\doteq$ 30.48 centimeters $4\frac{3}{4}$ inches $\doteq$ _____ centimeters.

THE EVOLUTION OF THE METRIC SYSTEM

The development of the metric system by the French Academy of Sciences was not universally applauded around the world. The United States, for example, continued to use the old English standards. In 1799, Congress noted that customs officers at every port seemed to have different opinions on the size of a pound or the length of a foot. Congress did pass a bill directing the establishment of a uniform standard for weights and measures, but never funded the bill to ensure its implementation.

By 1815, the need for a uniform unit of length caused the United States to order an 82 inch brass bar from the British. This bar, called the Troughton bar, was marked off in inches and the distance between the 27th and 63rd inch marks was designated the official unit of one yard. This was the unit used by the Coast and Geodetic Survey in all of its map-making work.

In 1838, following twenty years of study of the problems of measurement, a new set of standards for the United States were adopted. As before, these new standards ignored the metric system and were based on the old English system. The value of the new standards was limited to the fact that everybody in the country was now expected to use the common units. Those 1838 units were:

yard	The Troughton Bar
pound	Troy pound of England
gallon	231 cubic inches, which was the Queen Anne's wine gallon
bushel	2150.42 cubic inches, which was called the Winchester bushel

In 1893 the United States abolished the system of 1838 and redefined the meaning of yard and pound in terms of metric standards. The definitions of the 1893 yard and pound were:

$$1 \text{ yard} = \frac{3600}{3937} \text{ meter}$$

$$1 \text{ pound Avoirdupois} = \frac{1}{2.2046} \text{ kilogram}$$

These definitions were an improvement because they related the units of the United States to the metric system. However, the arithmetic involved in changing from one system to another discouraged much interchange and no other improvements were made to bring measuring systems together until recent years.

In Europe the metric system's efficiency became known and appreciated in business, manufacturing, and science. Many nations became interested in adopting the system. In 1870, the French government held an international conference which led to the signing of the "Metric Convention." During the convention an International Bureau of Weights and Measures was established which constructed new standards for the meter and kilogram. However, the scientists were still not satisfied with using a platinum bar as the physical representation of the meter. The length of the bar varied with its temperature. A natural, universal, and uniform standard was needed that could be reproduced almost anywhere, thereby eliminating the need to carry the basic standards from place to place.

In 1960, a natural standard for the meter was defined; this natural standard was one that could be easily reproduced in any well-equipped physics laboratory. The meter was defined as 1 650 763.73 wavelengths of the orange-red radiation of krypton 86. This new definition of the meter eliminated the chaos that could occur if the platinum bar that had been the meter standard was destroyed. The 1960 convention accomplished more than setting a universal natural basis for meter length. It reorganized the metric system into the Systeme International d'Unites, abbreviated SI. The SI established six basic measurement units. The six units and their SI names with abbreviations are given below.

Length	meter (m)
Mass	kilogram (kg)
Time	second (s)
Electric current	ampere (A)
Temperature	kelvin (K)
Luminous intensity	candela (cd)

Besides the six units listed above, the SI includes many other measurement units which are derived from the six basic units. Derived units are used to measure force, power, energy, and others. For uniformity and clarity, punctuation rules for the metric system were adopted. The punctuation rules for the metric system as adopted by the Systeme Internationale d'Unites are listed below.

1. The decimal point may be shown as on the following numbers:

   3.04 or 3·04 or 3,04

2. A number should never commence with a decimal point. $\frac{25}{100}$ should be written as 0.25, not as .25.

3. A number consisting of many digits should be arranged in groups of three with a full space and not a comma separating them.

   4 396 127.555 3 is the correct form not 4,396,127.5553

4. Abbreviations are to be written in singular form without a period. For example, km not kms. for kilometers.

One of the outstanding features of the SI system is the fact that multiples or divisions of all the basic units except the kilogram can be named just by attaching the correct prefix to the unit. For example, the unit that is 1 000 000 times longer than a meter is called a megameter and the unit that is one billion times smaller than the meter is called the nanometer. The metric prefixes and their associated multiplication factors are given below.

Prefix	Means
tera	one trillion times
giga	one billion times
mega	one million times
kilo	one thousand times
hecto	one hundred times
deca	ten times
deci	one tenth of
centi	one hundreth of
milli	one thousandth of
micro	one millionth of
nano	one billionth of
pico	one trillionth of

Since 1960, efforts have been made for adoption of the metric system in the United States. In 1971, Congress authorized a three year study of the feasibility of adopting the metric system. The report recommended that the metric system be adopted as the official system of weights and measures. In 1975, a bill providing for the voluntary and gradual conversion to the metric system in the United States was passed by the House and sent to the Senate. The law establishes a 21-member U. S. Metric Board that is responsible for planning the conversion, coordinating metric efforts already started, and conducting public eduation programs on metric measurement.

Meanwhile, improvements have been made in the metric system. In the future all basic units of measurement will be derived from natural constants such as the wave length of krypton 86. The kilogram will be obsolete. Measurements of mass (weight) will be made using some electrical or magnetic property of the electron or proton. Many other units will also be directly related to the properties of atoms, molecules, or crystals. It is expected that measurement instruments such as scales and meters will not be sent to a national standards laboratory for checking. Instead, a standards laboratory for checking instruments will be maintained at every factory or research facility. Local calibrations will be made automatically by a computer and related to the national standards. The wise man who said, "the only thing constant is change" appears to be right.

Progress Test 5

1. Which is longer, a megameter or a micrometer?

2. a. Name the metric linear unit that is one thousand times shorter than a meter.

   b. Name the metric unit one thousand times longer than a meter.

3. Which is shorter, a centimeter or a millimeter?

4. Write the following numerals using SI punctuation rules:

   a. 3492 centimeters b. .45 kms.

   c. 34,295.950043 mtrs. d. .0695 millim.

5. Place in order from the shortest to the longest. Millimeter, decameter, kilometer, micrometer, terameter, centimeter.

6. The gram is a unit of weight.

   a. Name the metric unit one thousand times larger than a gram.

   b. Name the unit one thousand times smaller than a gram.

THE RELATIVE SIZES OF METRIC UNITS

The first task of this section is to describe what is meant by a linear unit. A linear measurement is the measure of something "straight", such as a board or the length of a road. In inch-pound units, examples of linear measures are 6 feet, 13 miles, 2 inches, and 47 yards. In metric units, examples of linear measures are 3 centimeters, 40 kilometers, 97 millimeters, and 2 meters.

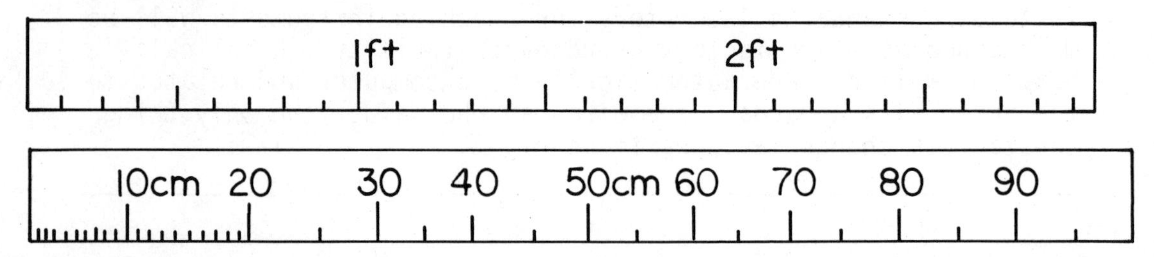

The drawings above show a reduced picture of a meter stick and a yard stick. Even though they are about $\frac{1}{4}$ their actual size, there are several lessons to be learned from these drawings.

The drawings show that a meter is slightly longer than a yard. A meter is approximately $39\frac{3}{8}$ inches long and a yard is 36 inches long.

The drawings also show that both the meter and the yard are subdivided into smaller units. The yard is divided into feet and inches. A meter (m) is divided into centimeters (cm) and the centimeters into smaller units called millimeters (mm). Notice that the larger units on the meter stick are labeled 10, 20, 30, and so on to 100. These units are called "centimeters". Therefore, 1 meter = 100 centimeters. Each centimeter (cm)is divided into ten smaller units about the thickness of a finger nail. Each of these smaller units is called a "millimeter" (mm). There are 10 millimeters in every centimeter. 1 cm = 10 mm. If a person counted the number of millimeters from the zero end of the meter stick to the 10 centimeter line, the count would be 10 x 10 millimeters or 100 millimeters. Similarly, 50 centimeters, which is half the length of a meter, is equivalent to 500 millimeters because 10 millimeters is equivalent to 1 centimeter and 50 x 10 mm = 500 mm. Since 1 meter is 100 centimeters long and 1 centimeter is 10 millimeters long,

1 meter = 100 x 10 millimeters

1 m = 1 000 mm

The "micron" is a metric linear measure that does not show on the meter stick because of its small size. The micron is a useful unit of measure for microscopic specimens because it is one-millionth of a meter long. Its small length allows micro-biologists to measure parts of micro-organisms using whole numbers.

The number of microns in a millimeter can be found using the fact that 1 meter is equivalent to 1 000 000 microns and also equivalent to 1 000 millimeters. As a result, 1 000 000 microns is equivalent in length to 1 000 millimeters.

1 000 000 microns = 1 000 mm

Dividing each side of the equality by 1 000 gives the number of microns in 1 millimeter.

1 000 microns = 1 mm

If the reader can imagine each of the 1 000 millimeters on a meter stick divided into 1 000 equal parts then each one of those parts would be 1 micron in length.

Progress Test 6

1. Place in order from the shortest length to the longest length. 1 meter, 1 micron, 1 centimeter, 1 millimeter.

   Fill in the blanks. Use the meter stick as an aide.

2. a. 1 meter is equivalent to _____ centimeters.

   b. 1 centimeter is equivalent to _____ meter.

   c. 1 meter is equivalent to _____ millimeters.

   d. 1 meter is equivalent to _____ microns.

   e. 1 millimeter is equivalent to _____ meter.

3. Which is longer, a 100 m race or a 100 yard race?

4. Since a meter is about $3\frac{3}{8}$ inches longer than a yard, a person 6 feet tall is _____(more, less) than 2 meters tall.

CONVERTING LINEAR UNITS: METRIC TO METRIC

Frequently, the solution of a problem in chemistry or physics requires measurements to be converted to other measurement units. A typical measurement conversion problem is given below.

2 000 millimeters = ? centimeters

The problem above asks how many centimeters (cm) are needed to have the same length as 2 000 millimeters. On the left is the label "millimeters" and on the right is the label "centimeters." In other words, the problem seeks to convert millimeters to centimeters. This can be done by dividing 2 000 millimeters (mm) by 10 millimeters. The justification for this division is the fact that 1 cm = 10 mm. The conversion can be made using the steps below.

$$\frac{2\ 000\ \text{mm}}{1} \times \frac{1\ \text{cm}}{10\ \text{mm}} = \frac{2\ 000}{1} \times \frac{1\ \text{cm}}{10} = \frac{2\ 000\ \text{cm}}{10} = 200\ \text{cm}$$

Two questions may arise at this point. Is the length of 2 000 millimeters equal to the length of 200 centimeters? If these measurements <u>are</u> equivalent, why did dividing by 10 millimeters and multiplying by 1 centimeter produce an equivalent measure? The answer to the first question is yes because 1 centimeter is equivalent to 10 millimeters so 200 centimeters is equivalent to 200 x (10 millimeters) or 2 000 millimeters. The answer to the second question is that dividing by 10 millimeters and multiplying by 1 centimeter produced an equivalent measure because 10 millimeters is equivalent to 1 centimeter. Consequently, $\frac{1\ \text{centimeter}}{10\ \text{millimeters}} = 1$. Since the fraction $\frac{1\ \text{centimeter}}{10\ \text{millimeters}}$ is equivalent to the number one (1), dividing 2 000 millimeters by 10 millimeters and multiplying by 1 centimeter is equivalent to multiplying 2 000 millimeters by the number 1. Therefore, an equivalent measure is produced because multiplying by 1 has no effect on the length of the measure but it does change the units.

Two uses of the number 1 are important to the conversion problems of this section.

1. For every number replacement of n, $1 \cdot n = n \cdot 1 = n$. Multiplying any number by 1 produces the same number.

2. 100 centimeters is equivalent in length to 1 meter. Therefore,

$$\frac{100 \text{ centimeters}}{1 \text{ meter}} = 1 \text{ and } \frac{1 \text{ meter}}{100 \text{ centimeters}} = 1.$$

Below another conversion problem is illustrated. The problem is to convert a measurement of 20 meters into centimeters. In other words, eliminate the label "meters" and obtain the label "centimeters". This change can be accomplished by multiplying by the numeral $\frac{100 \text{ centimeters}}{1 \text{ meter}}$ as shown in the computation below.

$$20 \text{ meters} = 20 \text{ meters} \cdot 1 = 20 \cancel{\text{meters}} \cdot \frac{100 \text{ centimeters}}{1 \cancel{\text{meter}}} =$$

2 000 centimeters. Notice that the fraction $\frac{100 \text{ centimeters}}{1 \text{ meter}}$ was used instead of the other numeral for 1, $\frac{1 \text{ meter}}{100 \text{ centimeters}}$, because the label "meter" was to be eliminated in the computation.

In the previous example, multiplying 20 meters by $\frac{100 \text{ centimeters}}{1 \text{ meter}}$ had the effect of eliminating the unit labeled "meters" because the result of the multiplication above was the measurement 2 000 centimeters. The elimination of other measurement units in the problems of this section will be accomplished similarly and the act of elimination will be called "cancelling out." In the next example, the microns will be "cancelled out."

268 microns is converted to meters by multiplying by the number 1 in the form of $\frac{1 \text{ meter}}{1\ 000\ 000 \text{ microns}}$ as shown in the following steps:

$$268 \text{ microns} = 268 \text{ microns} \cdot \frac{1 \text{ meter}}{1\ 000\ 000 \text{ microns}} = \frac{268}{1\ 000\ 000} \text{ meters}$$

or 0.000 268 meter.

To convert 25.4 kilometers (km) to meters, multiply 25.4 km by $\frac{1\ 000 \text{ m}}{1 \text{ km}}$ so that the label km will cancel out and the label m will be introduced.

$$25.4 \text{ km} = 25.4 \cancel{\text{km}} \cdot \frac{1\ 000 \text{ m}}{1 \cancel{\text{km}}} = 25\ 400 \text{ m}$$

The three previous examples show one method for converting metric measurements to other metric measurements using different units.

When given a measurement and asked to convert it to different units, the measurement will be multiplied by a numeral for 1. The numeral will be constructed so that the denominator will be in the same units as the measurement given and the numerator will be in the units desired.

Progress Test 7

1. $37 \text{ millimeters} = 37 \text{ millimeters} \cdot \frac{1 \text{ centimeter}}{10 \text{ millimeters}} = ?$

2. $25 \text{ centimeters} = 25 \text{ centimeters} \cdot \frac{10 \text{ millimeters}}{1 \text{ centimeter}} = ?$

3. $3.5 \text{ meters} = 3.5 \text{ meters} \cdot \frac{1 \text{ kilometer}}{1\ 000 \text{ meters}} = ?$

4. $2 \text{ kilometers} = 2 \text{ kilometers} \cdot \frac{1\ 000 \text{ meters}}{1 \text{ kilometer}} = ?$

5. $2\ 000\ 000 \text{ microns} = 2\ 000\ 000 \text{ microns} \cdot \frac{1 \text{ meter}}{1\ 000\ 000 \text{ microns}} = ?$

6. To change the measure, 7 meters, to centimeters which fraction,

   $\frac{1 \text{ meter}}{100 \text{ centimeters}}$ or $\frac{100 \text{ centimeters}}{1 \text{ meter}}$, should be used?

THE METRIC SYSTEM'S DECIMAL ADVANTAGES

The conversion advantages of the metric system's decimal organization will be discussed in this section.

The four examples of the previous section have shown the following statements to be true:

2 000 mm = 200 cm	268 microns = 0.000 268 m
25.4 km = 25 400 m	20 m = 2 000 m

The lesson to be learned from studying the examples above is that each conversion could have been made by moving the decimal point in the measurement correctly. The decimal point in 200 centimeters is one place to the left of its place in 2 000 millimeters.

2 000 millimeters can be converted to centimeters by moving the decimal point one place to the left to give 200.0 centimeters because 10 mm = 1 cm. 10 contains one zero and the decimal should be moved one place.

The decimal point in 20 meters is to the right of the unit's zero. The decimal point in 2 000 centimeters is two places right of its place in 20 meters. 100 cm = 1 m. 100 contains two zeros.

The decimal point in 0.000 268 meter is six places left of its place in 268 microns and 1 meter = 1 000 000 microns. 1 000 000 has six zeros.

The decimal point in 25 400 meters is three places right of its position in 25.4 kilometers and 1 000 meters = 1 kilometer. 1 000 has three zeros.

To convert 45 meters to kilometers, the decimal point in 45 should be moved three places to the left because meters are much shorter than kilometers and 1 000 meters = 1 kilometer. 1 000 contains three zeros. 45 m = 0.045 km.

To convert 25.4 centimeters to meters, the decimal point in 25.4 should be moved two places left because 1 meter = 100 centimeters and 100 contains two zeros. 25.4 = 0.254 m.

Progress Test 8

Convert the following measurements by moving the decimal point correctly.

1. 1 000 millimeters = 1 meter and 1 000 contains three zeros.

   Therefore, 2 574 mm = _____ m

2. 1 m = 100 cm and 100 contains two zeros. Consequently,

   5 839 cm = _____ m.

3. 1 000 m = 1 km and 1 000 contains three zeros. Therefore,

   2.5 km = _____ m

4. 3.6 m = _____ mm

5. 0.125 m = ______ mm

6. 0.5 m = ______ cm

Exercise Set 2

I. 1. Write the following numerals using the punctuation rules for SI.

a. 3049 cms. b. .58 kms. c. 3049.0084 gms.

d. 1,294 liters

2. Why is it more advantageous to use the wave length of the radiation of krypton 86 than a section of the arc of a meridian as a standard for the length of the meter?

3. Why is a metal bar not satisfactory as a standard for the meter?

4. Which of the following are not written correctly according to the punctuation rules for SI units:

a. 3 223 cm b. 4.559 35 1. c. 3,445 mms.

d. 4.988 3 cm e. .35 mm

5. Give the size of the following units in relation to the meter:

a. terameter b. kilometer c. megameter

d. hectometer e. centimeter f. decameter

g. micrometer h. millimeter

6. Place the following units in order from the smallest to the largest.

a. meter b. centimeter c. kilometer

d. inch e. yard f. millimeter

g. mile h. terameter i. micrometer

j. foot

7. a. 1 meter = _____ millimeters

   b. 1 meter = _____ kilometer

   c. 1 meter = _____ centimeters

   d. 1 meter = _____ micrometers

   e. 1 centimeter = _____ meter

   f. 1 kilometer = _____ meters

   g. 1 micrometer = _____ meter

   h. 1 millimeter = _____ meter

8. A person 6 feet tall is slightly less than 2 meters tall. 2 meters = _____ centimeters = _____ millimeters.

9. 20 miles is approximately 32 kilometers.

   32 km = _____ m = _____ cm.

10. 100 kilometers is approximately 62 miles.

    100 kilometers = _____ megameters = _____ meters.

11. $\frac{3}{4}$ inch is about 19 millimeters. 19 mm = _____ cm.

12. 13 mm is about $\frac{1}{2}$ inch. 13 mm = _____ microns = _____ cm.

13. $\frac{1\ 000 \text{ microns}}{1 \text{ mm}}$ is one numeral for 1 (one) that can be constructed from the statement 1 mm = 1 000 microns.

    Find another numeral for 1 that can be formed from that statement.

14. Write two numerals for one using the information in the statement 1 kilometer = 1 000 meters.

15. Fill in the blanks correctly.

    a. 25 m = _____ cm

    b. 450 cm = _____ m

    c. 35 km = _____ m

    d. 1 296.4 mm = _____ cm

e. 34 958 m = _____ cm

f. 384 823.384 m = _____ km

g. 12 microns = _____ mm

h. 2 m = _____ microns

i. 0.025 m = _____ microns

j. 0.5 kilometers = _____ meters

k. 0.125 cm = _____ mm

l. 0.025 6 km = _____ m

16. A lab technician wants to change the measure 0.005 7 cm to microns. 0.005 7 cm = _____ microns.

17. Find the equivalent measure for each of the following measures by moving the decimal point correctly:

a. 245 cm = _____ m

b. 0.000 006 m = _____ microns

c. 0.025 m = _____ cm

d. 35 840 km = _____ m

e. 35 microns = _____ mm

f. 35.4 m = _____ km

g. 2 384.385 m = _____ microns

h. 0.006 7 m = _____ cm

i. 274 cm = _____ mm

j. 45.9 mm = _____ cm

II. Challenge Problems

1. In 1893 the United States adopted the following definitions of the yard and pound:

$$1 \text{ yard} = \frac{3600}{3937} \text{ meter} \qquad 1 \text{ pound} = \frac{1}{2.204\ 6} \text{ kilogram}$$

Using the preceding definition, a. 1 meter = _____ yards and b. 1 kilogram = _____ pounds.

2. Fill in the blanks correctly.

   a. 47.6 terameters = _____ kilometers

   b. 0.045 8 gigameters = _____ millimeters

   c. 3 445 000 394.56 nanometers = _____ gigameters

   d. 0.000 000 35 decameters = _____ picometers

3. Which is longer, 0.000 045 6 megameters or 456 000 microns?

COMPLEX CONVERSIONS

Sometimes it is desirable to use two conversion facts to change a measurement to different units. For example, to answer the following problem it may be desirable to use the facts that 100 centimeters = 1 meter and 1 000 meters = 1 kilometer.

$$328 \text{ centimeters} = ? \text{ kilometers}$$

Since the meter is the basic linear metric unit, such conversions are often accomplished by first changing centimeters to meters and then changing meters to kilometers. Consequently, the conversion facts in the first paragraph above can be used to show the relation between centimeters and meters and then between meters and kilometers. First, cancel out the label "centimeter" by multiplying 328 cm by one as shown in the example below.

$$328 \text{ centimeters} = 328 \cancel{\text{centimeters}} \cdot \frac{1 \text{ meter}}{100 \cancel{\text{centimeters}}} = \frac{328}{100} \text{ meters}$$

The step above has eliminated the label "centimeters" but it also introduced the label "meter." The unit "meter" will be canceled out by multiplying by another numeral for the number one, $\frac{1 \text{ km}}{1\ 000 \text{ m}}$. The multiplication is shown below.

$$328 \text{ cm} = 328 \cancel{\text{cm}} \cdot \frac{1 \cancel{\text{m}}}{100 \cancel{\text{cm}}} \cdot \frac{1 \text{ km}}{1\ 000 \cancel{\text{m}}} = 0.003\ 28 \text{ km}$$

328 centimeters is equivalent to 0.003 28 kilometer

Since there are 100 cm in a meter and 1 000 m in a kilometer, there are 100 000 cm in a kilometer. This fact provides another way of converting 328 centimeters into kilometers.

$$328 \text{ cm} \cdot \frac{1 \text{ km}}{100\ 000 \text{ cm}} = \frac{328}{100\ 000} \text{ kilometers} = 0.003\ 28 \text{ km}$$

The next example converts a measurement in centimeters to microns. No conversion fact showing the relation between centimeters and microns has been given in this module, but the relation between microns and meters is known and the relation between meters and centimeters is known. Notice how these two facts are used to find the equivalent measure.

Sally, a laboratory technician measures a specimen and finds that it is 2.56 centimeters long. Using the fact that 100 centimeters = 1 meter and 1 meter = 1 000 000 microns, she converts the measure from centimeters to microns using the following steps:

$$\begin{aligned} 2.56 \text{ centimeters} &= 2.56 \cancel{\text{cm}} \cdot \frac{1 \cancel{\text{m}}}{100 \cancel{\text{cm}}} \cdot \frac{1\ 000\ 000 \text{ microns}}{1 \cancel{\text{meter}}} \\ &= \frac{2.56 \cdot 1\ 000\ 000}{100} \text{ microns} \\ &= 2.56 \cdot 10\ 000 \text{ microns} \\ &= 25\ 600 \text{ microns} \end{aligned}$$

One lesson to be learned from the previous example is the centimeter label and the meter labels canceled out because they were placed in the denominator and the microns label was introduced and then retained in the answer because it was in the numerator.

In general, one method of finding equivalent units is to change conversion facts into numerals for the number 1 arranged in such a way as to cancel out unwanted units and introduce the desired unit into the numerator when the measurement is multiplied by the numeral. The measurement can be multiplied by as many numerals for one as necessary to introduce the desired unit.

Progress Test 9

1. $48 \text{ kilometers} = 48 \text{ kilometers} \cdot \frac{1\ 000 \text{ meters}}{1 \text{ kilometer}} \cdot \frac{100 \text{ centimeters}}{1 \text{ meter}} = ?$

2. 150 centimeters = 150 centimeters $\cdot \dfrac{1\ \text{meter}}{100\ \text{centimeters}} \cdot \dfrac{1\ \text{kilometer}}{1\ 000\ \text{meters}} = ?$

3. Use the following conversion facts to change 0.025 kilometer to centimeters:

   1 kilometer = 1 000 meters

   100 centimeters = 1 meter

4. Use the following conversion facts to change 285 millimeters to decimeters:

   1 000 millimeters = 1 meter

   1 meter = 10 decimeters

CONVERTING UNITS OF MASS: METRIC TO METRIC

In the Systeme Internationale d'Unites, 1 kilogram, the basic standard of mass, is the platinum-iridium cylinder kept by the International Bureau of Weights and Measures in Paris. It weighs approximately 2.2 pounds. Since the kilogram is the only basic SI standard not measured from a natural phenomenon, the reader may be wondering why the standard for the kilogram was set at approximately 2.2 pounds. Why not 2 pounds, 3 pounds or some other convenient whole number? Fixing the weight of one kilogram at 2.2 pounds enabled units of weight to be related to units of volume. The French scientists, perhaps following the example of the Egyptians in 2000 BC, set the weight of one gram (a unit of mass) equal to the weight of one cubic centimeter (a unit of volume) of water, thereby establishing a relation between units of mass and units of volume. Volume units are derived from linear units, consequently, establishing 1 gram as the weight of 1 cubic centimeter (cc) of water at a certain temperature related mass and linear units.

1 gram (gm) = weight of 1 cubic centimeter (cc) of water

The following list shows the relations between the gram and other SI units of mass:

1 milligram = 0.001 gram

1 centigram = 0.01 gram

1 decigram = 0.1 gram

1 decagram = 10 grams

1 hectogram = 100 grams

1 kilogram = 1 000 grams

The number of centigrams equivalent to 14.5 hectograms can be found using the fact that 1 gm = 100 cgm and 1 hgm = 100 gm.

$$14.5 \text{ hgm} = 14.5 \text{ hgm} \cdot \frac{100 \text{ gm}}{1 \text{ hgm}} \cdot \frac{100 \text{ cgm}}{1 \text{ gm}} = 145\ 000 \text{ cgm}$$

14.5 hgm = 145 000 cgm

The decimal in 145 000 is four places right of its position in 14.5.

Progress Test 10

1. 1 gram = _____ kgm

2. 1 gram = _____ decigrams

3. 1 gram = _____ milligrams

4. A metric ton weighs 1 000 kilograms which is equivalent to 2 204.62 avoirdupois pounds. Find the number of grams equivalent to one metric ton.

5. A mechanic, working on a foreign car, needs to grind 18 milligrams from a crankshaft counterbalance. How many grams should he remove?

6. A pharmacist has to divide 1.5 kilograms of aspirin into pills that weigh five grams apiece. How many pills will he obtain?

METRIC LIQUID MEASURES

Only two metric units are used to measure liquids; the liter and the milliliter. Milk, gasoline, paint, and cooking oil are all sold by the liter. Medicines and other small amounts of liquids are sold by the milliliter so that whole numbers can be used.

1 liter = 1 000 milliliters

1 milliliter = 1 cubic centimeter

Reasoning from the two statements above, it can be said that 1 liter equals 1 000 cubic centimeters. Since 1 cubic centimeter of water weights 1 gram, 1 liter of water weighs 1 000 grams, or 1 kilogram. The fact that 1 cc of water weighs 1 gm enables the weight of a volume of water to be quickly calculated.

2.5 liters of water occupies 2 500 cubic centimeters of space and weighs 2 500 grams or 2.5 kilograms.

On the other hand, 25 ml of water weighs 25 gm or 0.025 kg. If a bottle will hold 454 gm of water, then its volume is 454 cubic centimeters or 0.454 liter.

Progress Test 11

1. A pharmacist finds that there is 0.045 ℓ of iodine in a bottle. Not wanting to give the amount as a fraction, the pharmacist gives the amount in milliliters. 0.045 ℓ = ? mℓ

2. 2 000 bottles of medicine each contain 50 ml of solution. How many liters of solution is there altogether?

3 One gallon of water weighs approximately 3 782 grams. How many liters of water would be equivalent to one gallon of water?

4. 20 gallons of gasoline weighs about 160 pounds and is equivalent to about 75 liters of gasoline. 20 gallons of gasoline would be _______ milliliters.

Exercise Set 3

I. 1. Ms. Mathy decided to write the measurement 0.005 6 meter in terms of microns so that it could be given in whole numbers. 0.005 6 m = ________ microns.

2. The measure 2 584 000 centimeters can be written as kilometers to reduce the number of digits in the measure. 2 584 000 cm = ____________ km .

3. Fill in the blanks correctly.

   a. 24 km = ________ cm

   b. 12 mm = ________ km

   c. 2 390.3 cm = ________ km

   d. 0.000 48 mm = ________ cm

   e. 0.23 km = ________ mm

   f. 2 395 microns = ________ mm

   g. 0.12 microns = ________ cm

   h. 0.05 cm = ________ microns

   i. 0.045 mm = ________ microns

   j. 3 485 003 microns = ________ km

4. 454 grams is approximately 1 pound. 454 gm = _____ kgm

5. A 350 pound defensive end weighs approximately 159 kilograms (kilos). 159 kilos = _______ milligrams.

6. 5 pounds of hamburger weighs about 2 200 grams. 2 200 gm = ________ decigrams.

7. A 12 ounce can of beverage will weigh about 340 grams. 340 gm = ________ kgm.

8. A 2 ounce fishing weight weighs about 56 grams. 56 gm = ________ mgm.

9. Fill in the blanks correctly.

   a. 45 kgm = __________ gm

   b. 0.5 kgm = __________ gm

   c. 0.125 kgm = __________ mgm

   d. 1.24 gm = __________ kgm

   e. 2 395 gm = __________ kgm

   f. 2 003 483 mgm = __________ gm

   g. 0.000 5 mgm = __________ kgm

   h. 48 040 990 mgm = __________ kgm

10. Fill in the blanks correctly.

   a. 2 000 mℓ = _____ ℓ

   b. 0.008 mℓ = _____ ℓ

   c. 34 cc = _____ mℓ

   d. 23 1 = _____ mℓ

   e. 0.5 ℓ = _____ cc

   f. 0.6 ℓ = _____ mℓ

   g. 34.5ℓ = _____ mℓ

   h. 0.034 085 ℓ = _____ mℓ

11. 150 portions of a beverage each contain 25 milliliters.

   a. 25 mℓ = _____ ℓ

   b. 25 ml of water = _____ grams

12. 100 gallons of oil is about 555 liters.

   a. 555 ℓ = _____ mℓ

   b. 555 ℓ of water = _____ gm of water

13. One-half gallon of milk is equivalent to about 2.27 liters of milk. 2.27 liters = _____ cubic centimeters.

14. 50 liters of gasoline is approximately 13 gallons of gasoline. 50 liters = _____ mℓ.

15. 1 gallon of wine is equivalent to approximately 454 milliliters of wine. 454 mℓ = _____ℓ.

16. a. Place in order from the smallest to the largest:

    1 kilogram, 1 centigram, 1 nanogram,

    1 decagram, 1 microgram, 1 milligram,

    1 gigagram, 1 gram

    b. Place in order from the largest to the smallest:

    1 milliliter, 1 kiloliter, 1 centiliter,

    1 liter, 1 hectoliter

    c. Which unit is out of place in the following list:

    1 kilometer, 1 milliliter, 1 centimeter,

    1 decameter, 1 micron

    d. Place in order from the smallest to the largest:

    256 centimeters, 25.6 meters, 25 600 000 millimeters,

    0.256 kilometers, 2 560 000 000 microns

II. Challenge Problems

1. Use the fact 1 gallon of water weighs about 8.337 pounds which is equivalent to approximately 3 782 grams. Find the kilograms of water in a swimming pool that contains 20 000 gallons of water.

2. Use the fact that 1 gallon of water has a volume of 231 cubic inches to find the kilograms of water that can be put in a tank 8 inches wide, 28 inches long and 12 inches high.

3. A water tank is 8 inches wide, 1 foot 3 inches long, and 2 meters 25 centimeters high. Find:

    a. The capacity of the tank in gallons, and

    b. the weight of the water in the tank in kilograms.

METRIC—INCH-POUND CONVERSIONS

One of the most difficult adjustments people will have to face if the United States adopts the metric system will be to "think metric" -- to use metric units for the old familiar measurements of the American culture: the weight of a baby, a man's height, or the volume of a gasoline tank.

During the changeover period from inch-pound to metric, conversion tables and devices will be widely used. However, each of the conversions can be accomplished by a relatively simple arithmetical computation. In this section, the computations involved in changing inch-pound measures to metric measures are explained.

Many babies are born approximately 21 inches long. The number of centimeters equivalent to 21 inches can be found by using the fact that 1 inch $\doteq$ 0.025 4 m. The symbol "$\doteq$" means "approximately equal." The number of centimeters equivalent to 21 inches can be found by multiplying both sides of 1 inch $\doteq$ 0.025 4 m by 21.

$$1 \text{ in.} \doteq 0.025\,4 \text{ m}$$

$$21 \cdot 1 \text{ in.} \doteq 21 \cdot 0.025\,4 \text{ m}$$

$$21 \text{ in.} \doteq 0.533\,4 \text{ m}$$

The number of centimeters equivalent to 21 inches can be found by correctly moving the decimal point in 0.533 4 m.

$$21 \text{ in.} \doteq 0.533\,4 \text{ m} = 53.34 \text{ cm} \doteq 53 \text{ cm}$$

Measurements are never exact. Consequently, it is more appropriate to state that 21 inches $\doteq$ 53.34 centimeters. No computed measurement can be more exact than the measurement from which it was derived.

Another familiar distance is the classic 100 yard dash. The unit "yard" is 3 feet long and is commonly used to measure items such

---

In 1959, Australia, Canada, New Zealand, South Africa, the United Kingdom, and the United States officially agreed to set the inch as exactly equal to 2.54 centimeters. The U. S. Coast and Geodetic Survey was allowed to keep its inch equal to 2.540 005 centimeters. Other inches in relation to the U. S. inch are El Salvador's pulgada, 0.913; Turkey's parmak, 1.24; and Swaziland's cape inch, 1.033.

as rugs and track events. The metric unit that will replace the yard is the meter. The meter is about $3\frac{3}{8}$ inches longer than a yard. Consequently, 1 yard $\doteq$ 0.914 4 meter.

$$100 \cdot 1 \text{ yard} \doteq 100 \cdot 0.914\ 4 \text{ meter}$$

$$100 \text{ yards} \doteq 91.44 \text{ meters} \doteq 91 \text{ meters}$$

The distance across the United States is approximately 3,000 miles. Will this distance given in kilometers be greater or less that 3,000? Since a kilometer is about 3 300 feet or $\frac{5}{8}$ of a mile, the distance across the United States is more than 3 000 kilometers.

$$1 \text{ mile} \doteq 1\ 609 \text{ meters}$$

$$3\ 000 \cdot 1 \text{ mile} \doteq 3\ 000 \cdot 1\ 609 \text{ meters}$$

$$3\ 000 \text{ miles} \doteq 4\ 827\ 000 \text{ meters}$$

$$4\ 827\ 000 \text{ meters} = 4\ 827 \text{ kilometers}$$

Americans will have to internalize the idea that the distance across the United States is about 4 800 kilometers.

The most common conversion facts relating inch-pound linear measures to meters are given below.

1 inch $\doteq$ 0.025 4 meter	or	1 meter $\doteq$ 39.37 inches
1 foot $\doteq$ 0.305 meter	or	1 meter $\doteq$ 3.281 feet
1 yard $\doteq$ 0.914 4 meter	or	1 meter $\doteq$ 1.093 yards
1 mile $\doteq$ 1 609 meter	or	1 meter $\doteq$ 0.000 621 mile

The number of inches equivalent to 36 centimeters can be found by multiplying both sides of the equivalence, 1 meter $\doteq$ 39.37 inches, by 0.36 because 36 cm = 0.36 m.

$$0.36 \cdot 1 \text{ meter} \doteq 0.36 \cdot 39.37 \text{ inches}$$

$$0.36 \text{ meter} \doteq 14.173\ 2 \text{ inches}$$

$$36 \text{ cm} \doteq 14.173\ 2 \text{ inches} \doteq 14 \text{ inches}$$

The number of miles equivalent to 80 kilometers can be found by multiplying both sides of the conversion fact 1 meter ≐ 0.000 621 mile by 80 000 because 80 km = 80 000 m.

80 000 · 1 meter ≐ 80 000 · 0.000 621 mile

80 000 meters ≐ 49.68 miles

80 kilometers ≐ 49.68 miles

80 kilometers ≐ 50 miles

Progress Test 12

1. 440 yards ≐ _____ meters.

2. 55 miles per hour ≐ _____ kilometers per hour.

3. The figures 36 - 26 - 36 inches have much meaning in American culture. When converted to centimeters the three previous numbers would read _____ - _____ - _____.

4. A 440 foot home run will be a _____ meter home run.

5. Kathy contemplates buying a suitcase 68 cm long to take on a trip 1 528 kilometers long.

   a. 68 cm ≐ _____ inches.

   b. 1 528 km ≐ _____ miles.

6. The term "furlong" is a popular inch-pound unit used usually in horse racing. 1 furlong = 220 yards. A bettor will need to think of races in terms of meters instead of furlongs.

   10 furlongs ≐ _____ meters.

METRIC—INCH-POUND MASS CONVERSIONS

Part of "thinking metric" will be to think kilograms instead of pounds and to think grams instead of ounces. Those who deal in large quantities will have to think of metric tons instead of tons. 1 metric ton = 1 000 kilograms.

George can find his 132 pound weight in kilograms using the conversion fact 1 pound ≐ 0.453 6 kilogram.

$$132 \cdot 1 \text{ pound} \doteq 132 \cdot 0.453\,6 \text{ kilogram}$$

$$132 \text{ pounds} \doteq 60 \text{ kilograms rounded off to the nearest kilogram.}$$

When the metric system is adopted in the United States, cold drinks may be sold in 341 gram bottles. The number of ounces in 341 grams can be found using the conversion fact 1 gram ≐ 0.035 2 ounce.

$$341 \cdot 1 \text{ gram} \doteq 341 \cdot 0.035\,2 \text{ ounce}$$

$$341 \text{ grams} \doteq 12 \text{ ounces}$$

Some of the conversion facts for measurement of mass are given below.

1 ounce ≐ 28.35 grams	1 gram ≐ 0.035 2 ounce
1 pound ≐ 0.453 6 kilogram	1 kilogram ≐ 2.204 6 pounds
1 ton ≐ 907 kilograms	1 metric ton ≐ 2 204.62 pounds

Progress Test 13

1. In the metric system a 300 pound defensive tackle will weigh _____ kilograms.

2. A 98 pound weakling will weigh _____ kilograms.

3. How many grams of hamburger will be needed to be equivalent to a four ounce patty?

4. A freshman has a blind date with a girl and finds out that she weighs 119 kilos. What is her weight in pounds?

5. If sugar is sold in 900 gram packages, how many pounds of sugar is in a package?

6. A 20 ton truck will hold _____ metric tons.

METRIC—INCH-POUND AREA CONVERSIONS

With respect to area, "thinking metric" will mean dealing with square meters instead of square yards, thinking of hectares instead of acres and using square centimeters instead of square inches.

If a family decides to install 40 square yards of carpet in its home, there may be a need to know the number of square meters of carpet required instead. The relation 1 square yard ≐ 0.836 square meter can be used to find the number of square meters of rug needed.

$$40 \cdot 1 \text{ square yard} \doteq 40 \cdot 0.836 \text{ square meter}$$

$$40 \text{ square yards} \doteq 33.44 \text{ square meters}$$

33 square meters has about the same area as 40 square yards because a meter is slightly longer than a yard.

The hectare is the French unit of land area equaling 10 000 square meters. 1 acre ≐ 0.405 hectare. Many farms in the United States contain 640 acres because there are 640 acres in a square mile. The number of hectares in 640 acres can be found using the steps below.

$$640 \cdot 1 \text{ acres} \doteq 640 \cdot 0.405 \text{ hectare}$$

$$640 \text{ acres} \doteq 259.2 \text{ hectares}$$

There are about 259 hectares on a 640 acre farm and there are about 259 hectares in a square mile.

Some area conversion facts are listed below.

1 square inch ≐ 6.45 square centimeters

1 square foot ≐ 0.092 square meter

1 square yard ≐ 0.836 square meter

An Englishman, Edmund Gunter, defined the "chain" as a unit of linear measure 66 feet long. The chain was used in the survey of the public lands of the United States. It is a convenient unit for the measure of land because a rectangle 1 chain wide and 10 chains long encloses exactly 1 acre.

1 square mile ≐ 2.59 square kilometers

1 acre ≐ 0.405 hectare

1 square centimeter ≐ 0.155 square inch

1 square meter ≐ 10.764 square feet or 1.196 square yards

1 square kilometer ≐ 0.386 square mile

1 hectare = 10 000 square meters ≐ 2.471 acres

A designer finds that she can buy 15 square meters of carpet for $45. How many square yards of carpet will she be able to buy for $45.?

The previous question can be more easily answered using the conversion fact 1 square meter = 1.196 square yards rather than the fact 1 square yard ≐ 0.836 square meter.

$$15 \cdot 1 \text{ square meter} \doteq 15 \cdot 1.196 \text{ square yards}$$

$$15 \text{ square meters} \doteq 17.94 \text{ square yards} \doteq 18 \text{ square yards}$$

Progress Test 14

1. In the "good old days" many people dreamed of having a farm consisting of 20 acres and a mule. 20 acres = ________ hectares.

2. A rectangular rug 15 feet by 30 feet contains 50 square yards. If the rug is priced by the square meter, approximately how many square meters are equivalent to 50 square yards?

3. A week-end sailor finds an ad in the paper for a sail containing 16 square meters of cloth and wonders how many square feet would make up 16 square meters. 16 square meters ≐ ________ square feet.

4. Kermit Remraf sold his 1 000 acre farm and plans to buy one containing 900 hectares. 900 hectares ≐ ________ acres.

5. If a bedspread measures 7 feet by 5 feet then its area is about ________ square meters.

METRIC—INCH-POUND LIQUID CONVERSIONS

In the inch-pound system liquids are measured using the following units.

1 gallon = 4 quarts

1 quart = 2 pints

1 pint = 2 cups

In the metric system liquids are measured in liters and milliliters. A liter is slightly larger than a quart. Adoption of the metric system will mean that milk, cream, whiskey, and other liquids will be sold by the liter instead of by the quart. Other smaller amounts of liquids such as medicines will be bought by the milliliter.

To find the number of liters that would make up 2 quarts the conversion fact 1 quart $\doteq$ 0.9463 liter can be used.

$$2 \cdot 1 \text{ quart} \doteq 2 \cdot 0.9463 \text{ liter}$$

$$2 \text{ quarts} \doteq 1.9 \text{ liters rounded to the nearest tenth}$$

A person buying 2 gallons of milk would have to buy 2 · 3.785 liters or 7.57 liters of milk. 1 gallon $\doteq$ 3.785 liters.

The number of milliliters equivalent to one-half a cup of milk can be found using the conversion fact 1 cup $\doteq$ 0.236 5 liters.

$$0.5 \cdot 1 \text{ cup} \doteq 0.5 \cdot 0.236\ 6 \text{ liters}$$

$$0.5 \text{ cup} \doteq 0.118\ 3 \text{ liters or } 118.3 \text{ milliliters}$$

$$0.5 \text{ cup} \doteq 100 \text{ milliliters}$$

The "fifth" is a popular bottle size used for alcoholic beverages. Actually it contains one-fifth of a gallon or four-fifths of a quart. By coincidence it is the same size as the British quart -- the quart used in England before adoption of the metric system. After metric adoption the fifth may still be popular as it is about three-fourths of a liter.

The number of liters and kilograms of water in a swimming pool 15 feet by 30 feet by 8 feet deep can be found by first finding its dimensions in meters as shown below. 1 foot ≐ 0.304 8 meter.

15 feet ≐ 15 · 0.304 8 meter = 4.572 meters

30 feet ≐ 30 · 0.304 8 meter = 9.144 meters

8 feet ≐ 8 · 0.304 8 meter = 2.438 4 meters

The volume of the pool is found by multiplying the length times the width times the depth.

Volume = 4.572 · 9.144 · 2.438 4 = 102 cubic meters rounded to the nearest cubic meter. Since 1 cubic meter contains 1 000 000 cubic centimeters, 1 cubic meter contains 1 000 000 milliliters. Consequently, 102 cubic meters contain 102 000 000 milliliters of water which weighs 102 000 000 grams or 102 000 kilograms. Remember 1 milliliter of water weighs 1 gram.

Some conversion facts concerning liquids are given below.

1 cup ≐ 0.236 6 liter	1 liter ≐ 4.224 cups
1 pint ≐ 0.473 1 liter	1 liter ≐ 2.112 pints
1 quart ≐ 0.946 3 liter	1 liter ≐ 1.056 quarts
1 gallon ≐ 3.785 liters	1 liter ≐ 0.264 gallon

Progress Test 15

1. Five gallons of gasoline ≐ ______ liters of gasoline.

2. 3 quarts of milk ≐ ______ liters of milk.

3. A 16 ounce bottle ≐ ______ milliliters of liquid.

4. If a watering trough is a rectangular shape 10 feet long, 2 feet wide and 1 foot deep, then it contains about ______ liters of water which weigh ______ kilograms.

5. If cream is sold in 0.7 liter containers, approximately how many pints of cream are in one of the containers?

6. Mr. Reyub buying an oil tank finds that it holds 100 liters. Approximately how many gallons of oil can he put in the tank?

TEMPERATURE CONVERSIONS: KELVIN, CELSIUS, FAHRENHEIT

In 1954, the unit of temperature, the Kelvin (K) was added to the metric system as a basic unit. The Kelvin was then adopted by the Systeme Internationale d'Unites as one of the six basic units.

The Kelvin temperature scale consists of only positive numbers as the temperature at which molecular activity ceases was assigned the lowest point, zero, on the Kelvin scale. The temperature 273.16^0 K is the triple point of water: the temperature at which water can exist in a liquid, gaseous or solid state. 273.16^0 K is slightly above the freezing point of water as 273.15^0 K is the freezing point of water. The Kelvin scale is usually used by scientists in metric countries.

The Celsius scale is used for other temperature measurements. The International Bureau of Weights and Measures renamed the Centigrade scale for Celsius, the Swedish astronomer who invented it. 273.16 K is equivalent to 0 degrees on the Celsius scale and 32 degrees on the Fahrenheit scale. The three temperature scales are compared in the figure below.

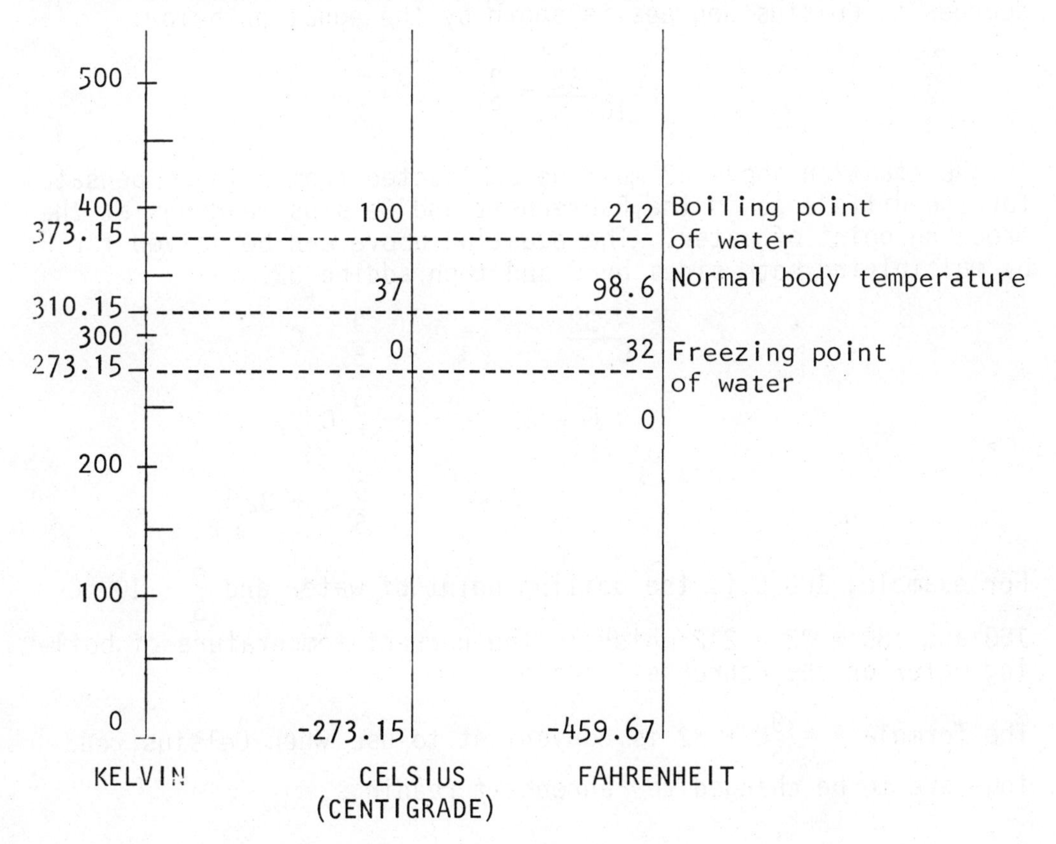

There are several lessons to be learned from the figure above.

1. The size of the units on the Kelvin scale and the Celsius scale are equal. There are 100 degrees difference between the boiling point of water and the freezing point of water on both scales.

2. The Kelvin degrees are the same size as the Celsius degrees, but the Celsius zero is 273 degrees above the Kelvin zero. Therefore, a Celsius reading is the same temperature as a Kelvin reading minus 273 degrees. The difference in the two scales, 0.15 K will be ignored in these computations. $C = K - 273$ or $K = C + 273$.

3. On the Fahrenheit scale there is 180 degrees difference between the freezing point and the boiling point of water. $212 - 32 = 180$ degrees.

Therefore, the ratio of Fahrenheit degrees to Celsius degrees is

$$\frac{212 - 32}{100} = \frac{180}{100} = \frac{9}{5}$$

Using F to represent the number of Fahrenheit degrees and C to represent the number of Celsius degrees, the ratio of Fahrenheit degrees to Celsius degrees is shown by the equation below.

$$\frac{F - 32}{1C} = \frac{9}{5}$$

In the equation above 32 must be subtracted from F to compensate for the difference in the Fahrenheit and Celsius readings at the freezing point of water. The equation above can be solved for F by multiplying both sides by C and then adding 32.

$$C \cdot \frac{F - 32}{C} = \frac{9}{5} \cdot C$$

$$F - 32 = \frac{9}{5} C$$

$$F = \frac{9}{5} C + 32$$

For example, 100^0C is the boiling point of water and $\frac{9}{5} \cdot 100^0C =$ 180 and $180 + 32 = 212$ which is the correct temperature of boiling water on the Fahrenheit scale.

The formula $F = \frac{9}{5}C + 32$ is convenient to use when Celsius readings are to be changed to Fahrenheit readings.

To change 37^0C to F, the following steps can be used:

$$F = \frac{9}{5}C + 32$$

$$F = \frac{9}{5} \cdot 37 + 32$$

$$F = \frac{333}{5} + 32$$

$$F = 66.6 + 32$$

$$F = 98.6^0$$

$$37^0C = 98.6^0F$$

37^0C is the normal temperature of the human body.

To change Fahrenheit readings to Celsius a different formula can be used. $C = \frac{5}{9}(F - 32)$ is the convenient formula to use if Fahrenheit readings are to be converted to Celsius readings. Notice that 32 has to be subtracted from the Fahrenheit reading before the multiplier $\frac{5}{9}$ is used because the Celsius temperature for the freezing point of water is 32 degrees below the equivalent Fahrenheit figure.

The reading 212^0F can be converted to the equivalent Celsius reading by replacing F with 212 in the formula $C = \frac{5}{9}(F - 32)$ and using the following steps:

$$C = \frac{5}{9}(F - 32)$$

$$C = \frac{5}{9}(212 - 32)$$

$$C = \frac{900}{9}$$

$$C = 100^0$$

Two more examples of converting Fahrenheit readings to Celsius readings and of converting Celsius readings to Fahrenheit readings will be given.

The temperature sometimes falls as low as -70^0F at the North Pole. -70^0F can be converted to Celsius degrees by replacing F with

-70 in the formula $C = \frac{5}{9}(F - 32)$ and using the steps below.

$$C = \frac{5}{9}(-70 - 32)$$

$$C = \frac{5}{9} \cdot -102$$

$$C = \frac{-510}{9}$$

$$C = -56.6$$

The temperature in a blast furnace for melting steel is read by two thermometers; one a Celsius scale and the other a Fahrenheit scale. To find the correct reading for the Fahrenheit thermometer when the Celsius thermometer reads 458 degrees, C can be replaced with 458 in the formula $F = \frac{9}{5}C + 32$.

$$F = \frac{9}{5} \cdot 458 + 32$$

$$F = \frac{4122}{5} + 32$$

$$F = 824.4 + 32$$

$$F = 856.4$$

The equation $F = \frac{9}{5}C + 32$ can be solved for C using the following steps:

$$F = \frac{9}{5}C + 32$$

$$F - 32 = (\frac{9}{5}C + 32) - 32$$

$$F - 32 = \frac{9}{5}C$$

$$\frac{5}{9}(F - 32) = \frac{5}{9} \cdot \frac{9}{5}C$$

$$\frac{5}{9}(F - 32) = 1 \cdot C$$

$$C = \frac{5}{9}(F - 32)$$

Consequently, the two formulas $F = \frac{9}{5}C + 32$ and $C = \frac{5}{9}(F - 32)$ are called equivalent equations; they have the same solutions.

Progress Test 16

1. What is the reading on a Celsius thermometer when a Kelvin thermometer in the same room reads 300 degrees?

2. 72^0 F is a comfortable temperature for a summer day. What is the equivalent reading on a Celsius scale?

3. 38 degrees sounds pretty cold to Americans because they are used to the Fahrenheit temperature scale. 38^0C = ? F.

4. -40^0C = ? F.

WHEN WILL THE UNITED STATES ADOPT THE METRIC SYSTEM?

The main roadblocks to the adoption of the metric system in the United States are the costs of conversion and the resistance of many citizens. There is no doubt that there will be many expenses involved when the United States adopts the metric system. However, when the conversion is spread over a long period of time the cost may seem negligible when compared to the money saved by the metric system's simplicity. Many products now in use that were built and calibrated to inch-pound units can be replaced when they are worn out with metric oriented replacements.

The resistance of many people to a new measuring system may actually be a more difficult problem in the long run than the cost of metric conversion. Some of the people in the countries that adopted the metric system many years ago still use old familiar units. Florists in Paris trying to sell roses by tens instead of by the dozen encountered stiff buyer resistance and had to go back to selling roses by the dozen. The metal plates especially made for serving raw oysters still have 12 instead of 10 depressions in some Paris restaurants.

Even the advocates of the metric system experience difficulties "thinking metric." At one of the meetings of the Metric Association, a group that had been preaching the advantages of the metric system since 1915, the president, exhibiting a meter stick to the group called it a "metric yardstick." The mind of the president was still "thinking inch-pound" as there is no metric yard stick -- the yard is not a metric unit.

Exercise Set 4

I. 1. When the metric system is adopted, the 880 yard run may be measured in meters. 880 yards $\doteq$ __________ meters.

2. 1 500 meters is the length of an olympic race. 1 500 m $\doteq$ ____________ yards.

3. How many feet long is a 5 000 meter race?

4. The common concrete block is 8 in. by 8 in. by 16 in. long. Find its dimensions in centimeters.

5. The yard will be replaced by the meter when the metric system is adopted. 12 m $\doteq$ __________ yards.

6. 80 kilometers $\doteq$ __________ miles.

7. 200 miles per hour $\doteq$ _________ kilometers per hour.

8. When the metric system is adopted, centimeters will replace inches. Measure your waist, feet, fingers, biceps, head size and other body parts using centimeters.

9. a. 12 in. $\doteq$ _____ cm   d. 30 cm $\doteq$ _____ in.

   b. 36 in. $\doteq$ _____ cm   e. 122 cm $\doteq$ _____ in.

   c. 60 in. $\doteq$ _____ cm   f. 183 cm $\doteq$ _____ in.

10. a. 30 mph $\doteq$ _____ kmph   c. 90 kph $\doteq$ _____ mph

    b. 45 mph $\doteq$ _____ kmph   d. 60 kph $\doteq$ _____ mph

11. When the metric system is adopted, grams will replace ounces, kilograms will replace pounds and metric tons will replace tons.

    a. 6 ounces of aspirin $\doteq$ _______ grams.

    b. 10 pounds $\doteq$ ____________ kilograms

c. 10 tons of sand weighs about ______ metric tons.

d. 15 kilos ≐ ______ pounds.

e. A 181 kilo center on a football team weighs about ______ pounds.

f. 250 grams ≐ ______ ounces.

g. 1 296.4 gm ≐ ______ pounds.

h. 0.35 pounds ≐ ______ gm

12. To "think metric" is to think of square centimeters instead of square inches and to think of hectares instead of acres.

a. 36 square inches ≐ ______ square centimeters.

b. There are 1 296 square inches in a square yard.
1 296 sq. in. ≐ ______ sq. cm

c. 3 square miles ≐ ______ square meters.

d. 100 acres ≐ ______ hectares

e. 20 square centimeters ≐ ______ square inches.

f. 25 square kilometers ≐ ______ square miles.

g. 0.45 hectares ≐ ______ acres.

h. 20 square meters ≐ ______ square yards.

i. 1 294.38 square centimeters ≐ ______ square feet.

13. Liters will be used instead of gallons and milliliters will be used instead of ounces when SI units are adopted.

a. 20 liters of oil ≐ ______ gallons of oil.

b. 32 mℓ ≐ ______ cups.

c. 8.0 ℓ ≐ ______ pints.

d. 4.0 liters ≐ ______ quarts

e. 8.0 quarts ≐ ______ liters.

f. 100 gallons ≐ ______ liters

g. 0.5 gallon $\doteq$ ________ milliliters.

h. 0.5 cup of milk $\doteq$ ______ milliliters of milk.

14. In the metric system the Celsius scale is used for temperature measurement instead of the Fahrenheit scale.

a. $80^0F \doteq$ _____ ^0C.

b. 212^0F is the boiling point of water at sea level. $212^0F \doteq$ _____ C.

c. $0^0C \doteq$ _____ ^0F.

d. At sea level the freezing point of water is 32^0F. $32^0F \doteq$ _____ ^0C.

e. 90^0F is a typical summer temperature. $90^0F \doteq$ _____ ^0C.

f. $^-20^0C \doteq$ _____ ^0F.

g. Some times the temperature in the winter may go as low as $^-15^0F$. $^-15^0F \doteq$ _____ ^0C.

h. $^-70^0F$ in Alaska is about _____ ^0C.

15. a. Place in order from the largest to the smallest.

1 kilometer, 1 inch, 1 decameter, 1 yard,

1 centimeter, 1 foot, 1 millimeter, 1 mile,

1 micron, 1 meter

b. Place in order from the smallest to the largest.

1 pound, 1 gram, 1 ton, 1 kilogram, 1 ounce,

1 metric ton, 1 centigram

c. Place in order from the smallest to the largest.

1 liter, 1 cubic inch, 1 milliliter, 1 cubic foot,

1 cubic meter, 1 cubic yard, 1 cubic centimeter,

1 cubic millimeter

d. Place in order from the smallest to the largest.

1 acre, 1 square foot, 1 square mile, 1 hectare,

1 square meter, 1 square yard, 1 square inch,

1 square kilometer, 1 square centimeter

II. Challenge Problems

1. A stick 9 feet 4 inches long is about ______ meters long.

2. Find the volume in cubic meters of a box 12 feet 3 inches by 18 inches by 2.65 meters.

3. Find the number of kilograms of oil in a cylindrical tank 14 feet in diameter and 10 feet high if a gallon of oil weighs about 8 pounds.

4. 12.5 metric tons of aspirin is to be sold as 50 mg tablets for 2¢ each. Find the value of the aspirin in tablet form.

5. A painter uses paint that will cover 11 square meters per liter on a building 40 meters long, 30 meters wide and 12 meters high. Each side is a rectangle. The paint costs $2. a liter. Find the cost of one coat of paint.

6. Find the temperature at which a Fahrenheit thermometer and a Celsius thermometer will have the same reading.

7. A farmer is to fertilize his land at the rate of 90 kilos per hectare. The fertilizer costs $3.50 per 45 kilogram sack and he has a field 750 meters by 1250 meters to fertilize. How much will the fertilizer cost?

APPENDIX

The definitions of the six basic SI units are listed below.

Length - Meter - m: The meter is defined as 1 650 763.73 wavelengths in vacuum of the orange-red line of the spectrum of krypton 86.

Mass - Kilogram - kg: The standard for the unit of mass is a cylinder of platinum-iridium alloy kept by the International Bureau of Weights and Measures in Paris.

Time - Second - s: The second is defined as the duration of 9 192 631 770 cycles of the radiation associated with a specified transition of the cesium atom.

Temperature - Kelvin - K: The thermodynamic or Kelvin scale of temperature has its origin or zero point at absolute zero. The temperature of the triple point of water (when water can exist in solid-liquid-gaseous states) is defined to be 273.16 kelvins.

Electric Current - Ampere - A: The ampere is defined as the magnitude of the current that, when flowing through each of two long parallel wires separated by one meter in free space, results in a force between the wires of 2×10^{-7} newton for each meter of length.

Luminous Intensity - Candela - cd: The candela is defined as the luminous intensity of $\frac{1}{600\ 000}$ of a square meter of a radiating cavity at the temperature of freezing platinum (2 042 K).

DERIVED UNITS

QUANTITY	UNIT	SYMBOL	
Area	square meter	m^2	
Volume	cubic meter	m^3	
Frequency	hertz	Hz	(s^{-1})
Density	kilogram per cubic meter	kg/m^3	
Velocity	meter per second	m/s	
Angular velocity	radian per second	rad/s	
Acceleration	meter per second squared	m/s^2	
Angular acceleration	radian per second squared	rad/s^2	
Force	newton	N	(kgm/s^2)
Pressure or stress	newton per square meter	N/m^2	
Kinematic viscosity	square meter per second	m^2/s	
Dynamic viscosity	newton-second per square meter	$N \cdot s/m^2$	
Work, energy, quantity of heat	joule	J	(N·m)
Power	watt	W	(J/s)
Electric charge	coulomb	C	(A·s)
Voltage, potential difference, electromotive force	volt	V	(W/A)
Electric field strength	volt per meter	V/m	
Electric resistance	ohm	Ω	(V/A)
Electric capacitance	farad	F	(A·s/V)
Magnetic flux	weber	Wb	(V·s)
Inductance	henry	H	(V·s/A)
Magnetic flux density	tesla	T	(Wb/m^2)
Magnetic field strength	ampere per meter	A/m	
Magnetomotive force	ampere	A	
Flux of light	lumen	lm	(cd·sr)
Luminance	candela per square meter	cd/m^2	
Illumination	lux	lx	(lm/m^2)

From: *Prepare Now For a Metric Future,* Donovan, F., Weybright and Talley, 1970

Basic Conversion Facts between the Metric and Inch-Pound Systems

Length

1 meter	$\doteq$ 1.093 yards	1 yard	$\doteq$ 0.914 4 meter
	$\doteq$ 3.281 feet	1 foot	= 30.48 centimeters
	$\doteq$ 39.37 inches	1 inch	= 2.54 centimeters
1 kilometer	$\doteq$ 0.621 miles	1 mile	$\doteq$ 1.609 kilometers

Area

1 square meter $\doteq$ 1.196 square yards
$\doteq$ 10.764 square feet
1 square centimeter $\doteq$ 0.155 square inch
1 square kilometer $\doteq$ 0.386 square mile
1 hectare $\doteq$ 2.471 acres
1 square yard $\doteq$ 0.836 square meter
1 square foot $\doteq$ 0.092 square meter
1 square inch $\doteq$ 6.45 square centimeters
1 square mile $\doteq$ 2.59 square kilometers
1 acre $\doteq$ 0.405 hectare

Volume

1 cubic meter $\doteq$ 1.308 cubic yards
1 cubic yard $\doteq$ 0.764 cubic meter
1 cubic centimeter $\doteq$ 0.61 cubic inch
1 cubic foot $\doteq$ 0.028 cubic meter
1 cubic inch $\doteq$ 16.387 cubic centimeters

Volume

1 liter	$\doteq$ 1.056 quarts	1 quart	$\doteq$ 0.946 liter
	$\doteq$ 0.264 gallon	1 gallon	$\doteq$ 3.785 liters

Mass

1 gram	$\doteq$ 15.432 grains	1 grain	$\doteq$ 0.064 9 gram
	$\doteq$ 0.035 2 ounce	1 ounce	$\doteq$ 28.35 grams
1 kilogram	$\doteq$ 2.204 6 pounds	1 pound	$\doteq$ 0.453 6 kilogram
1 metric ton	$\doteq$ 2 204.62 pounds	1 ton	$\doteq$ 0.907 metric ton

MODULE SELF-TEST

1. What difficulties are involved in using the length of the King's foot as a linear measurement unit?

2. Place the following in order from the smallest to the largest:

   1 centigram, 1 kilogram, 1 gram, 1 hectogram, 1 milligram

3. Which is the better measurement unit for weight? Why?

   a. The weight of a cupful of barley.

   b. The weight of a cupful of water.

4. Which is heavier?

   a. 1 metric ton or 1 English ton.

   b. 1 kilogram or 1 pound.

   c. 1 gram or 1 ounce.

5. Which is shorter?

   a. 1 meter or 1 yard.

   b. 1 centimeter or 1 inch.

   c. 1 mile or 1 kilometer.

6. Fill in the blanks correctly.

   a. 1 decimeter = _____ meter.

   b. 1 kilometer = _____ meter.

   c. 1 decameter = _____ meter.

   d. 1 centimeter = _____ meter.

7. Fill in the blanks correctly.

   a. 754 cm = _____ m

   b. 63.4 decimeters = _____ m

c. 450 grams = _____ kgm

d. 57 kgm = _____ mgm

e. 456 milliliters = _____ liters

8. Change each of the following to a different unit of measure using the formula shown:

a. 1 inch ≐ 2.54 cm   17 inches ≐ _____ cm

b. 1 yard ≐ 0.914 4 m or 1 m ≐ 1.093 yards.
540 yards ≐ _____ m

c. 1 liter ≐ 1.056 quarts or 1 quart ≐ 0.946 3 liter
12 quarts ≐ _____ liters

d. $57^{0}F$ ≐ _____ ^{0}C, $C = \frac{5}{9}(F - 32)$ or $F = \frac{9}{5}C + 32$

e. 100 liters ≐ _____ gallons, 1ℓ ≐ 0.264 gallon or
1 gallon ≐ 3.785ℓ

f. 1 gram ≐ 0.035 2 ounce   536 gm ≐ _____ pounds

g. 1 m ≐ 0.000 621 mile   55 kilometers ≐ _____ miles

h. 1 pound ≐ 0.453 6 kgm or 1 kgm ≐ 2.204 6 pounds
125 pounds ≐ _____ kilograms

READING LIST

How You Could Possibly Live With the Metric System, John E. Robinson, Jr., Human Design Advocates, 1972

SI Units, Chiswell and Grigg, John Wiley and Sons, Australasia Pty, Ltd., 1971

Prepare Now For a Metric Future, Frank Donovan, Weybright and Talley, 1970

Twentieth Yearbook of the National Council of Teachers of Mathematics, Bureau of Publications, Teachers College, Columbia University, 1948

Metric System, Le Maraic, Abbey Books, 1973

U. S. Metric Study Interim Report, U. S. National Bureau of Standards, Government Printing Office, 1971

Weights and Measures: An Informal Guide, Stacy U. Jones, Public Affairs Press, Washington, D.C., 1963

PROGRESS TEST ANSWERS

Progress Test 1

1. b 2. a 3. No answer

Progress Test 2

1. 1 foot = $\frac{2}{3}$ cubit. 2. Homer 3. a. 30 cups

b. 300 cups c. 3 000 cups d. true

Progress Test 3

1. 54 2. $2\frac{2}{3}$ 3. 512 4. 65 536

Progress Test 4

1. The length of the unit is not standardized. It varies from King to King.

2. millimeter, centimeter, meter, hectometer, kilometer

3. 1 000 4. 10 5. $\frac{1}{10}$

Progress Test 5

1. megameter 2. a. millimeter b. kilometer 3. millimeter
4. a. 3 492 cm b. 0.45 km c. 34 295.950 043 d. 0.069 5 mm
5. micrometer, millimeter, centimeter, decameter, kilometer, terameter
6. a. kilogram b. milligram

Progress Test 6

1. 1 micron, 1 millimeter, 1 centimeter, 1 meter

2. a. 100 b. $\frac{1}{100}$ c. 1 000 d. 1 000 000 e. $\frac{1}{1\ 000}$

3. 100 m race 4. less

Progress Test 7

1. 3.7 centimeters 2. 250 millimeters 3. 0.003 5 kilometer

4. 2 000 meters 5. 2 meters 6. $\frac{100 \text{ centimeters}}{1 \text{ meter}}$

Progress Test 8

1. 2.574 m 2. 58.39 m 3. 2 500 m 4. 3 600 mm
5. 125 mm 6. 50 cm

Progress Test 9

1. 4 800 000 cm 2. 0.001 5 km 3. 2 500 cm
4. 2.85 decimeters

Progress Test 10

1. 0.001 2. 10 3. 1 000 4. 1 000 000 5. 0.018
6. 300

Progress Test 11

1. 45 2. 100 3. 3.782 4. 75 000

Progress Test 12

1. 402 2. 88 3. 91 - 61 - 91 4. 134 5. a. 27
b. 949 6. 2 012

Progress Test 13

1. 136 2. 44 3. 113 4. 262 5. 1.98 6. 18

Progress Test 14

1. 8.1 2. 42 3. 172 4. 2 220 5. 3.2

Progress Test 15

1. 20 2. 2.8 3. 470 4. 566,566 5. 1.5 6. 26

Progress Test 16

1. 27^0C 2. 22^0C 3. 100^0F 4. $^-40^0$F

EXERCISE SET ANSWERS

Exercise Set 1

I. 1. a. 10 b. 0.001 c. 10 000 d. 2 500 e. 3.5
f. 457
2. a. 2.5 b. 6 c. 45 d. 76 e. 2.54 f. 343
3. a. 3 b. 4 c. 150 d. $\frac{1}{2}$ e. 5 f. $14\frac{1}{2}$
g. 75 h. $\frac{3}{2}$

Exercise Set 1 (continued)

4. a. 6 b. 24 c. $1\frac{1}{2}$ d. $1\frac{1}{10}$ e. Chief Highiq
5. The King had a short foot compared to the other rulers of his time.
6. a. 262 144 b. 192 c. 128
7. It is difficult to measure a portion of a meridian accurately.
8. a. 2 b. 2 c. 3 d. 5 000 e. 500 000
   f. 300 g. 3 h. 500 i. 0.2 j. 0.5
9 To change inch-pound units to other inch-pound units, the multipliers and divisors are not powers of 10. Changing units in the metric system only involves moving the decimal point.

II. 1. a. 0.025 b. 8.56 c. 34 950 000 d. 0.00002
2. 13.4 3. a. 480 b. 480 000 000 4. 12.065

Exercise Set 2

I. 1. a. 3 049 cm b. 0.58 km c. 3 049.008 4 gm
   d. 1 294ℓ
2. The wave length of Krypton 86 can be reproduced relatively easily compared to the measurement of a portion of a meridian.
3. The metal bar changes its length depending upon the temperature of the bar. Also, its length may change due to wear or damage.
4. b, c, e
5. a. one trillion meters. b. 1 000 meters
   c. one million meters d. one hundred meters
   e. 0.01 meter f. 10 meters g. one-millionth of a meter h. one-thousandth of a meter
6. micrometer, millimeter, centimeter, inch, foot, yard, meter, kilometer, mile, terameter

7. a. 1 000 b. 0.001 c. 100 d. 1 000 000
   e. 0.01 f. 1 000 g. 0.000 001 h. 0.001
8. 200, 2 000
9. 32 000, 3 200 000

10. 0.1, 100 000 11. 1.9 12. 13 000, 1.3
13. $\frac{1 \text{ mm}}{1\ 000 \text{ microns}}$ 14. $\frac{1 \text{ km}}{1\ 000 \text{ m}}$, $\frac{1\ 000 \text{ m}}{1 \text{ km}}$
15. a. 2 500 b. 4.5 c. 35 000 d. 129.64
    e. 3 495 800 f. 384.823 384 g. 0.012
    h. 2 000 000 i. 25 000 j. 500 k. 1.25 l. 25.6
16. 57

Exercise Set 2 (continued)

17. a. 2.45 b. 6 c. 2.5 d. 35 840 000 e. 35 000
    f. 0.035 4 g. 2 384 385 000 h. 0.67 i. 2 740
    j. 4.59

II. 1. a. 1.093 611 b. 2.204 6

2. a. 47 600 000 b. 45 800 000 000
   c. 0.000 000 003 445 000 394 56 d. 3 500 000

3. 0.000 045 6 megameters

Exercise Set 3

I. 1. 5 600 2. 25.84

3. a. 2 400 000 b. 0.000 012 c. 0.023 903
   d. 0.000 048 e. 230 000 f. 2.395 g. 0.000 012
   h. 500 i. 45 j. 0.003 485 003

4. 0.454 5. 159 000 000

6. 22 000 7. 0.34 8. 56 000

9. a. 45 000 b. 500 c. 125 000 d. 0.001 24
   e. 2.395 f. 2 003.483 g. 0.000 000 000 5
   h. 48.040 990

10. a. 2 b. 0.000 008 c. 34 d. 23 000 e. 500
    f. 600 g. 34 500 h. 34.085

11. a. 0.025 b. 25

12. a. 555 000 b. 555 000

13. 2 270 14. 50 000 15. 0.454

16. a. 1 nanogram, 1 microgram, 1 milligram, 1 centigram, 1 gram, 1 decagram, 1 kilogram, 1 gigagram

    b. 1 kiloliter, 1 hectoliter, 1 liter, 1 centiliter, 1 milliliter

    c. 1 milliliter

    d. 256 cm, 25.6 m, 0.256 km, 2 560 000 000 microns, 25 600 000 mm

Exercise Set 3 (continued)

II. 1. 75 640 kgm

2. 44

3. a. 46 b. 174 kgm

Exercise Set 4

I. 1. 805

2. 1 640 yards

3. 16 405

4. 20.32 cm by 40.64 cm by 20.32 cm

5. 13

6. 50

7. 322

8. No answer given

9. a. 30.48 b. 91.44 c. 152.4 d. 11.8
e. 48 f. 72

10. a. 48 b. 72 c. 56 d. 37

11. a. 170 b. 45 c. 9 d. 33 e. 399 f. 8.8
g. 2.858 h. 159

12. a. 232 b. 8 359 c. 7 770 000 d. 40.5
e. 3.1 f. 9.7 g. 1.1 h. 24 i. 1.393

13. a. 5.3 b. 0.14 c. 17 d. 4.2 e. 7.6
f. 379 g. 1 892 f. 115

14. a. 27^0C b. 100^0C c. 32^0F d. 0^0C e. 194^0F
f. $^-40^0$F g. $^-26^0$C h. $^-57^0$C

15. a. 1 mile, 1 kilometer, 1 decameter, 1 meter, 1 yard,
1 foot, 1 inch, 1 centimeter, 1 millimeter, 1 micron

Exercise Set 4 (continued)

b. 1 centigram, 1 gram, 1 ounce, 1 pound, 1 kilogram, 1 ton, 1 metric ton

c. 1 cubic millimeter, 1 milliliter = 1 cubic centimeter, 1 cubic inch, 1 liter, 1 cubic foot, 1 cubic yard, 1 cubic meter

d. 1 square centimeter, 1 square inch, 1 square foot, 1 square yard, 1 square meter, 1 acre, 1 hectare, 1 square kilometer, 1 square mile

II. 1. 2.8

2. 4.5

3. 41 804 kgm

4. $5 000 000

5. $306 for 153 liters of paint

6. $^-40^0$F = $^-40^0$C

7. $658 for 188 sacks

MODULE SELF-TEST ANSWERS

1. It is not easily accessible or standardized

2. 1 milligram, 1 centigram, 1 gram, 1 hectogram, 1 kilogram

3. The weight of a cupful of water because its weight will not change as easily as a cupful of barley.

4. a. 1 metric ton b. 1 kilogram c. 1 ounce

5. a. 1 yard b. 1 centimeter c. 1 kilometer

6. a. 0.1 b. 1 000 c. 10 d. 0.01

7. a. 7.54 b. 6.34 c. 0.450 d. 57 000 000 e. 0.456

8. a. 43 b. 494 c. 11.4 d. 14^{0}C e. 26.4 f. 1.18 g. 34 h. 56.7